LESSONS IN LINEAR ALGEBRA

LESSONS IN LINEAR ALGEBRA

James R. Wesson
Vanderbilt University

Charles E. Merrill Publishing Company
A Bell & Howell Company
Columbus, Ohio

MERRILL MATHEMATICS SERIES

Erwin Kleinfeld, *Editor*

Published by
Charles E. Merrill Publishing Co.
A Bell & Howell Company
Columbus, Ohio 43216

Copyright ©, 1974, by Bell & Howell Company. All rights reserved. No part of this book may be reproduced in any form, electronic or mechanical, including photocopy, recording, or any information storage or retrieval system, without permission in writing from the publisher.

International Standard Book Number: 0-675-08831-3

Library of Congress Catalog Number: 73-91403

1 2 3 4 5 6 — 79 78 77 76 75 74

Printed in the United States of America

To Jan and Janni

PREFACE

This text is designed to be used in a one-term course for students who have completed a two- or three-term course in calculus and analytic geometry. It is also assumed that the student has at least a slight acquaintance with geometric vectors in two- or three-space, but any deficiency here can be overcome with a little extra effort.

The Mathematical Association of America, through its Committee on the Undergraduate Program, has made recommendations as to how linear algebra could fit into different programs. Every recommendation includes as main topics (in a linear algebra course) systems of linear equations, matrices, vector spaces, linear transformations, and real Euclidean space, with an emphasis on two- and three-space. These topics are covered in this text.

The study of general linear spaces has been placed at the very beginning of the text. Although it may be more difficult for the student to start with this subject rather than with the rules of matrix algebra, I am convinced that the student will benefit from the integrity of this approach, whether he is a liberal arts student, social science student, scientist, or engineer. The student who goes on for further work in mathematics, computing science, or statistics will be well prepared to apply linear algebra to his field. The student who takes the course as his last one in mathematics will forget many details but will remember how an abstraction can be used in order to unify many problems into one.

The main part of a course is given in Lessons 1 through 18. I recommend that they be studied in sequence, even though Lessons 5, 9, 10 could be studied without the others as background. Needs of different students can be met by emphasizing different topics. For example, all students should be exposed to n-tuple spaces, but certain function spaces are more interesting to engineers than to social scientists. It may be appropriate to assign different exercises to different students.

Preface

Lessons 19 and 20 deal with various special applications of linear algebra. The lesson on preliminaries at the beginning of the text is intended as a reference.

It is not necessary that the student go through the proof of every theorem, and I see no reason for the student to memorize proofs. He may go through some and accept the others without proofs, but it is important that he knows when he is proving something and when he is assuming something. The Halmos symbol ∎ is used to denote the end of a proof, even if part is left for the student.

Definitions and most theorems are followed immediately with illustrative examples. The examples of the text may be reinforced by showing on the blackboard other examples, especially those which can be illustrated in two- or three-space. Answers to most exercises are given at the end of each lesson.

I acknowledge my debt to many authors whose works have influenced me. I especially mention Tom Apostol, Daniel T. Finkbeiner, P. R. Halmos, D. C. Murdoch, and Evar Nering — each the author of well-known texts suitable for a linear algebra course.

Especially I acknowledge the encouragement and suggestions by Erwin Kleinfeld and W. R. Hutton, Jr. as they examined the work for the Charles E. Merrill Publishing Company. Also I am appreciative of specific suggestions by Margaret Kleinfeld and E. B. Shanks as they examined prepublication copies, and I appreciate the help of editor Nancy Sander in preparing the manuscript for publication. I am grateful to the University of California at Berkeley for providing office space and access to library resources as I wrote the first draft of the text. Finally, I extend thanks to my wife for her typing and retyping of the manuscript.

CONTENTS

Preliminaries   1

Lesson 1
Definition and Examples of Linear Spaces (Vector Spaces)   15

Lesson 2
Subspaces, Linear Dependence and Linear Independence   21

Lesson 3
More on Subspaces and Linear Dependence — Bases, Dimension   30

Lesson 4
More on Bases, Spanning Sets, and Dimension   40

Lesson 5
Applications to Systems of Linear Equation   49

Lesson 6
Inner Products   60

Lesson 7
Orthogonality   70

Lesson 8
Orthogonal Complements   82

Lesson 9
Matrices and Matrix Operations   92

Lesson 10
Some Uses of Matrix Notation and Matrix Algebra   104

Contents

Lesson 11
Elementary Matrices, the Inverse of a Matrix, and Rank of a Matrix   118

Lesson 12
Linear Transformations   131

Lesson 13
Rank, Nullity, and Inverse of a Linear Transformation   145

Lesson 14
The Linear Problem — More Operations on Linear Transformations   157

Lesson 15
Determinants   170

Lesson 16
Applications of Determinants — Orthogonal Matrices and Isometries   184

Lesson 17
Eigenvalues and Eigenvectors   200

Lesson 18
Similarity and Canonical Forms   214

Lesson 19
Applications to Geometry   234

Lesson 20
Quadratic Forms and Quadric Surfaces   254

Bibliography   271

Index   273

PRELIMINARIES

The lessons which follow are intended for the student who has completed a two or three term course in calculus. It is assumed that you know the elementary facts about sets, functions or mappings, geometric vectors, analytic geometry, and calculus. If you have forgotten some of the details you may have to consult another book; however, these lessons are written so that they can be understood with very little reference to other sources.

This preliminary lesson will serve as a reference in case you need to refresh your memory on a notation or fact. Much of the summary merely reminds you of the notations we will use as we employ concepts that you already understand. Facts are not proved here, but most of them are illustrated. If you have time, you may enjoy sketching some of the proofs.

A reasonable approach would be for you to skim through this preliminary lesson, skipping the parts which you have already studied, even though you have forgotten some of the details. Treat this lesson as material to be consulted later when needed.

You might want to pursue further some topic discussed here. If so, you will enjoy the exercises which are intended to expose and illustrate the various topics of the lesson.

Sets and Notations

We generally use block letters A, B, ... to denote sets, and we follow commonly accepted notations.

Notation	Meaning
$\Rightarrow$	"Implies."
$x \in A$	x is an element of A, or x belongs to A.

Lessons in Linear Algebra

Notation	Meaning
$x \notin A$	x does not belong to A.
$\{x \in A \mid x \text{ has property } p\}$	The set of all elements x which belong to A and have property p.

(Remark: Whenever the context is clear, we may use the notation $\{x \mid x \text{ has property } p\}$.)

Notation	Meaning
$S \cap T$	The *intersection* of S and T, $\{x \mid x \in S \text{ and } x \in T\}$.
$S \cup T$	The *union* of S and T, $\{x \mid x \in S \text{ or } x \in T\}$.
$S \subset T$	S is a subset of T. That is, $x \in S \Rightarrow x \in T$.
$S = T$	$S \subset T$ and $T \subset S$.
$\emptyset$	The empty set.
$S - T$	The elements in S but not in T, $\{x \mid x \in S \text{ and } x \notin T\}$.
$\bigcap_{i \in I} S_i$	The intersection of all sets S_i, where i belongs to the "indexing" set I.
$\bigcup_{i \in I} S_i$	The union of all sets S_i, where $i \in I$.

To illustrate the above, let R be the set of all real numbers, and let S_i be the set of all real numbers in the closed interval $[i, i+1]$, that is, let $S_i = \{x \in R \mid i \leq x \leq i+1\}$. Then

$$S_{1/2} \cup S_0 = \left[0, \frac{3}{2}\right]$$

$$S_{1/2} \cap S_0 = \left[\frac{1}{2}, 1\right]$$

$$S_1 \cap S_3 = \emptyset$$

$$\bigcup_{i \in R} S_i = R$$

$$\bigcap_{i \in R} S_i = \emptyset$$

$$\bigcap_{i \in [0,1]} S_i = \{1\}$$

Mappings

A *mapping (function, transformation)* f of set A into set B is a rule which assigns to each element $x \varepsilon A$ a unique element $f(x) \varepsilon B$. Alternatively, a mapping f of A into B can be thought of as a set $\{(x, y) \mid x \varepsilon A, y \varepsilon B\}$ of ordered pairs such that if (x_1, y_1) and (x_2, y_2) are two ordered pairs with $x_1 = x_2$, then $y_1 = y_2$.

We will use the notation $f: A \rightarrow B$ to mean a mapping f of A into B. (See Figure P-1). If $x \varepsilon A$, then $f(x)$ is called the *image of x*, A is called

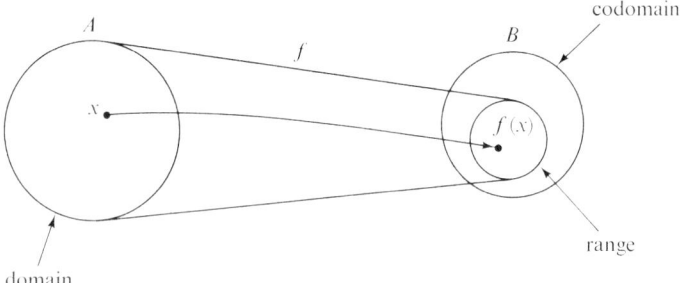

FIGURE P-1

the *domain* of f, B the *codomain* of f. The set of all images of A, denoted $f(A)$, is called the *range* of f. That is,

$$\text{range } f = \{y \varepsilon B \mid y = f(x) \text{ where } x \varepsilon A\}$$

Let y be an element in B. The *inverse image* of y, denoted $f^{-1}(y)$ is the set of all elements of A which are mapped into y. That is,

$$f^{-1}(y) = \{x \varepsilon A \mid f(x) = y\}$$

This set is also called the *complete inverse image* of y.

The mapping f is said to be *one-to-one* (or 1-1) if distinct elements of A have distinct images in B. This is the same as requiring for $x_1, x_2 \varepsilon A$,

$$x_1 \neq x_2 \Longrightarrow f(x_1) \neq f(x_2)$$

or

$$f(x_1) = f(x_2) \Longrightarrow x_1 = x_2$$

The mapping f is said to be a mapping of A *onto* B if each element of B is the image of at least one element of A, or

$$y \varepsilon B \Longrightarrow y = f(x) \quad \text{where } x \varepsilon A$$

Two mappings f and g with the same domain A are said to be *equal* if $f(x) = g(x)$ for all $x \varepsilon A$.

Lessons in Linear Algebra

The *composition* of mappings is the one used in calculus. For $f: A \to B$ and $g: B \to C$, we define $gf: A \to C$ by $(gf)(x) = g(f(x))$ for all $x \in A$. (This is illustrated in Figure P-2.) It is not difficult to show that composition of mappings is associative, in that $h(gf) = (hg)f$.

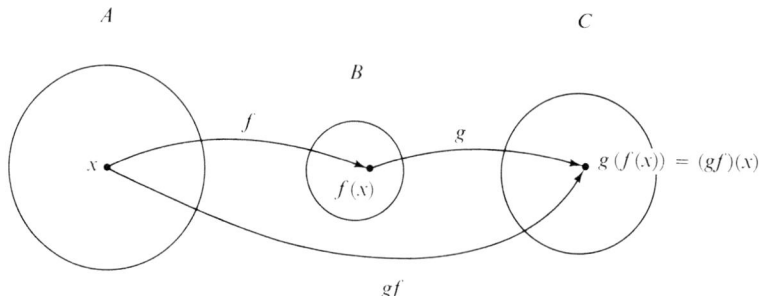

FIGURE P-2

The mapping which sends each element of A into itself is called the *identity mapping on* A, denoted 1_A or 1. This means $1(x) = x$ for all $x \in A$.

Let $f: A \to B$. If $g: B \to A$ such that $gf = 1_A$, then g is called a *left-inverse* of f. If $fg = 1_B$ then g is a *right-inverse* of f. If $gf = 1_A$ and $fg = 1_B$, then g is called the *inverse* of f.

The mapping $f: A \to B$ has a right-inverse if and only if f is onto. For if f is onto, define $g: B \to A$ by $g(y) = x$, where x is some element

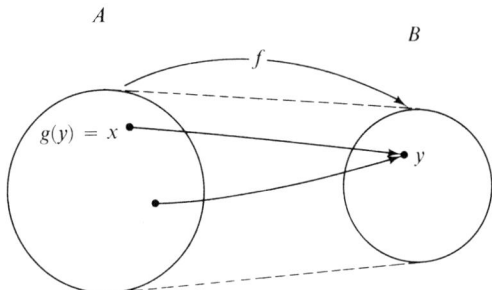

FIGURE P-3

for which $f(x) = y$ (x need not be unique). Then $(fg)(y) = f(x) = y$, and $fg = 1_B$.

Conversely, let $g: B \to A$ so that $fg = 1_B$. If $y \in B$, then $(fg)(y) = 1_B(y) = y$, and f maps $g(y)$ into y. This means that f is *onto*.

The mapping $f: A \to B$ has a left-inverse g if and only if f is 1-1.
The condition (necessary and sufficient) for f to have an *inverse* f^{-1} is that f be 1-1 and onto.

Example 1. Let A be the set of all integers $\{\ldots -2, -1, 0, 1, 2, \ldots\}$, and let f, g, h be mappings of $A \to A$, with $f(n) = n + 1$, $g(n) = n^2$,

$$h(n) = \begin{cases} n + 1 & \text{if } n \geq 0 \\ n & \text{if } n < 0 \end{cases}$$

Then f is 1-1 and onto, g is neither 1-1 nor onto, and h is 1-1 but not onto. The mapping f has an inverse $f^{-1}(n) = n - 1$, and h has a left-inverse

$$k(n) = \begin{cases} n - 1 & \text{if } n \geq 1 \\ n & \text{if } n < 1 \end{cases}$$

The range of g is the set of squares $\{0, 1, 4, 9, \ldots\}$. Under g, the inverse image of 4 is the set $\{2, -2\}$. Examples of composites are $(fg)(n) = n^2 + 1$, $(gf)(n) = (n+1)^2$.

The Summation Notation

The summation notation will be used freely throughout the book. The idea is simple, but to use it to advantage requires practice.

$$\sum_{i=1}^{n} f(i) \text{ means } f(1) + f(2) + \ldots + f(n)$$

For example,

$$\sum_{i=1}^{3} x_i = x_1 + x_2 + x_3$$

$$\sum_{j=1}^{4} 2^j = 2^1 + 2^2 + 2^3 + 2^4 = 30$$

$$\sum_{k=2}^{4} (k + \log k) = (2 + \log 2) + (3 + \log 3) + (4 + \log 4)$$

It is easy to recognize that

$$\sum_{i=1}^{n} f(i) = \sum_{j=1}^{n} f(j) = \sum_{i=3}^{n+2} f(i-2), \ldots$$

Also, we have distributive laws such as

$$\sum_{i=1}^{n} 2x_i = 2 \sum_{i=1}^{n} x_i$$

We will frequently use double summations, so you need to understand why

$$\sum_{i=1}^{m} \sum_{j=1}^{n} f(i, j) = \sum_{j=1}^{n} \sum_{i=1}^{m} f(i, j)$$

Occasionally, when the context is clear, we will write simply $\sum x_i$ to mean $\sum_{i=1}^{n} x_i$.

Basics from Analytic Geometry and Calculus

Your background in analytic geometry and calculus may be entirely adequate for this book. We will assume that you are familiar with the differentiation and integration of elementary functions, although you may have to review the details when they are needed. We will use facts such as:

If $f(x)$ is continuous on $[a, b]$, then $\int_a^b f(t)\,dt$ exists.

If f is differentiable at a, then f is continuous at a.

If $\int_a^x f(t)\,dt$ exists for all x in $[a, b]$, then $\dfrac{d}{dx} \int_a^x f(t)\,dt = f(x)$.

If f, g are integrable on $[a, b]$ and $f(x) \leq g(x)$ for all x in $[a, b]$, then $\int_a^b f(t)\,dt \leq \int_a^b g(t)\,dt$.

Often these kinds of facts can be recalled by a look at graphs.

In Figure P-4 rectangular coordinates systems in two-space and three-space are used.

In three-space the Pythagorean theorem gives the distance between $P_1(x_1, y_1, z_1)$ and $P_2(x_2, y_2, z_2)$ by $d^2 = (x_1 - x_2)^2 + (y_1 - y_2)^2 + (z_1 - z_2)^2$. A linear equation $ax + by + cz + d = 0$ represents a plane.

Preliminaries

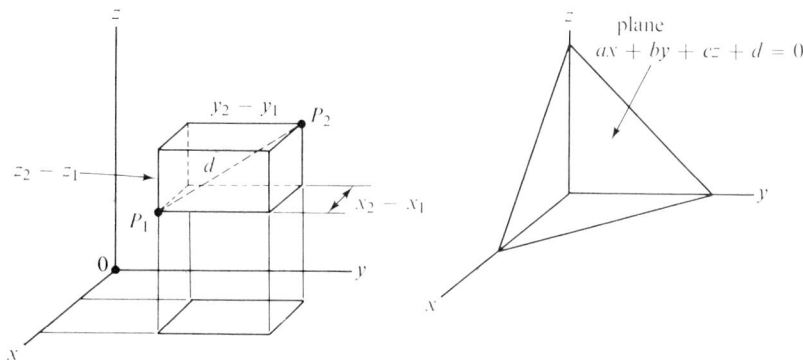

FIGURE P-4

In some applications a *vector* is defined as a quantity with direction and magnitude, while a scalar is taken as a quantity with just magnitude. Velocity is a vector, speed is a scalar. Force is a vector, weight is a scalar. A directed line segment (an arrow) is a vector, and it is customary to use arrows to symbolize other vectors. For example, an arrow 2 units long pointing north can represent a wind from the south of 20 miles per hour. An arrow pointing downward can represent the force of gravity on a man. Using arrows to represent vectors permits one to associate the length and direction of the arrow with the magnitude and direction of the vector.

In the majority of these lessons, the concept of a vector is much more general than that of a geometric vector. However, geometric vectors will remain as special cases of the generalized concept and will serve as meaningful and useful illustrations.

In space, the vector (arrow) from point A to point B is written $\overrightarrow{AB}$. Two vectors are *equal* if they have the same directions and the same lengths. Two vectors need not coincide to be equal; however, the equality defined is an *equivalence relation*. This means: (1) a vector equals itself; (2) if $\overrightarrow{AB} = \overrightarrow{CD}$, then $\overrightarrow{CD} = \overrightarrow{AB}$; (3) if $\overrightarrow{AB} = \overrightarrow{CD}$ and $\overrightarrow{CD} = \overrightarrow{EF}$, then $\overrightarrow{AB} = \overrightarrow{EF}$. Names for these properties are: (1) the *reflexive* property; (2) the *symmetric* property; (3) the *transitive* property.

We take the "line segment" from a point to itself as the *zero vector*, and the vector $-\overrightarrow{AB}$ is the vector having the same length as $\overrightarrow{AB}$ but the opposite direction. The *scalar multiple* $r\overrightarrow{AB}$ of a vector is the vector whose length is $|r|$ times that of $\overrightarrow{AB}$ and whose direction is the same as that of $\overrightarrow{AB}$ if $r > 0$ and opposite that of $\overrightarrow{AB}$ if $r < 0$. (See Figure P-5.)

Geometric Vectors

Lessons in Linear Algebra

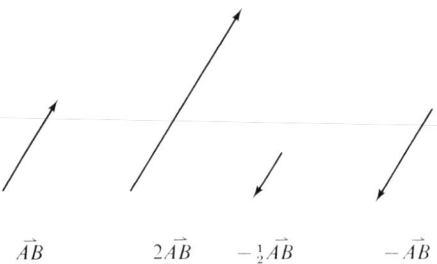

FIGURE P-5

Two vectors $\vec{AB}$ and $\vec{CD}$ are added as follows (see Figure P-6): Draw from B a vector $\vec{BE} = \vec{CD}$. Then $\vec{AB} + \vec{CD}$ is taken as $\vec{AE}$ The difference of two vectors is given by $\vec{AB} - \vec{CD} = \vec{AB} + (-\vec{CD})$.

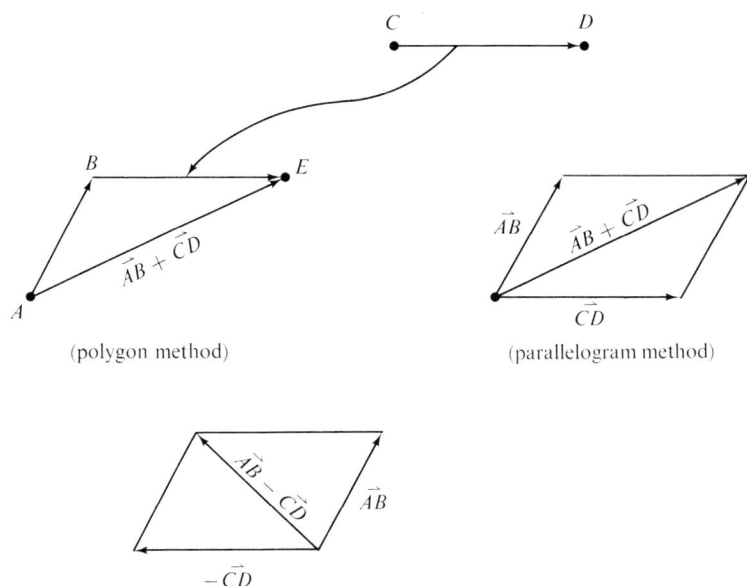

FIGURE P-6

It is not difficult to verify that vector algebra satisfies a number of laws. For example, vector addition is commutative and associative. These and other properties will be examined for more general vectors later.

If we represent each vector as an arrow beginning at the origin, then each vector in space can be represented by the coordinates of the end point. So in two-space, we occasionally represent geometric vectors as (x_1, y_1), (x_2, y_2), etc., and on three-space we have vectors (x_1, y_1, z_1), (x_2, y_2, z_2), etc. The addition defined by $(x_1, y_1) + (x_2, y_2) = (x_1 + x_2, y_1 + y_2)$ agrees with the preceding geometric definition of addition of vectors. Similarly, in three-space we have $(x_1, y_1, z_1) + (x_2, y_2, z_2) = (x_1 + x_2, y_1 + y_2, z_1 + z_2)$.

Often the "unit" vectors $(1, 0, 0)$, $(0, 1, 0)$, $(0, 0, 1)$ are referred to as $\vec{i}, \vec{j}, \vec{k}$, respectively. Then $(x, y, z) = x\vec{i} + y\vec{j} + z\vec{k}$.

The "dot product" of two geometric vectors (see Figure P-8) $\vec{OP_1}$ and $\vec{OP_2}$ is defined by $\vec{OP_1} \cdot \vec{OP_2} = |\vec{OP_1}| |\vec{OP_2}| \cos \theta$, the product of the lengths of the two vectors and the cosine of the included angle. It is easy to verify that $\vec{OP_1} \cdot \vec{OP_2} = x_1 x_2 + y_1 y_2$ in the plane and $\vec{OP_1} \cdot \vec{OP_2} = x_1 x_2 + y_1 y_2 + z_1 z_2$ in space.

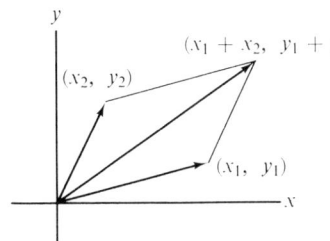
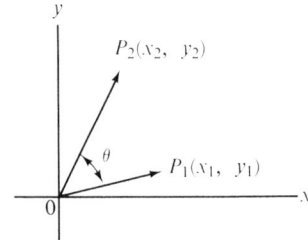

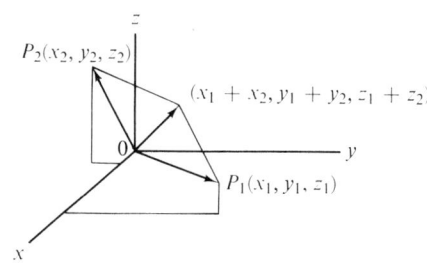

FIGURE P-7

Most of the lessons which follow deal with the field of real numbers, but optional material is included with parallel results for the complex number field. We summarize here basic facts and notations about complex numbers.

Let $z = a + bi$ be a complex number, with a, b real and $i^2 = -1$.

Complex Numbers

Lessons in Linear Algebra

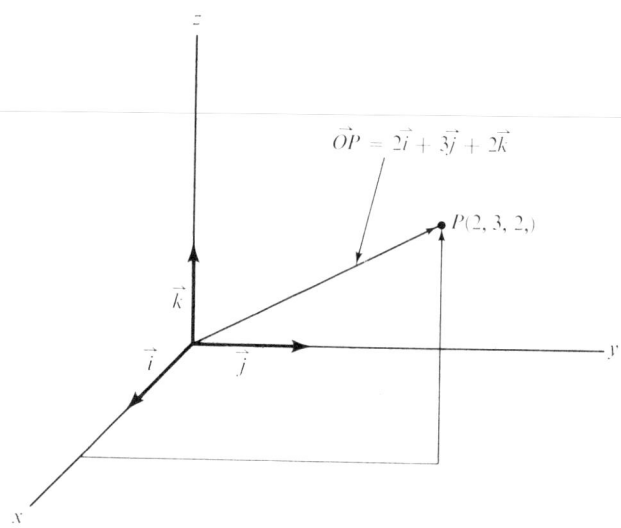

FIGURE P-8

The number z is considered *real* if $b = 0$, *imaginary* if $b \neq 0$, and *pure imaginary* if $a = 0$ and $b \neq 0$. The number a is called the *real part* of z, denoted $\text{Re}(z)$ or $\text{Real}(z)$. The *imaginary part* of z is given by $\text{Im}(z) = b$. The *conjugate* of z is denoted $\bar{z}$ and is defined by $\bar{z} = a - bi$. The *absolute value* (*modulus*) of z is denoted $|z|$ and is defined by $|z| = \sqrt{z \bar{z}}$.

The following properties are immediate for complex numbers z, z_1, and z_2.

$$\overline{z_1 + z_2} = \overline{z_1} + \overline{z_2}$$

$$\overline{z_1 z_2} = \overline{z_1}\,\overline{z_2}$$

$$\overline{\left(\frac{z_1}{z_2}\right)} = \frac{\overline{z_1}}{\overline{z_2}} \quad \text{if } z_2 \neq 0$$

$$|z_1|\,|z_2| = |z_1 z_2|$$

$$\left|\frac{z_1}{z_2}\right| = \frac{|z_1|}{|z_2|} \quad \text{if } z_2 \neq 0$$

$$z + \bar{z} = 2\text{Re}(z)$$

$$z - \bar{z} = 2i\,\text{Im}(z)$$

$$|\text{Re}(z)| \leq |z|$$

$$|z_1 + z_2| \leq |z_1| + |z_2|$$

Preliminaries

Example 2. Let $z_1 = 2 + 3i$, $z_2 = 3 - i$. Then $\bar{z}_1 = 2 - 3i$, $\bar{z}_2 = 3 + i$, $\overline{z_1 + z_2} = \overline{5 + 2i} = 5 - 2i$, $z_1 z_2 = 9 + 7i$, $\bar{z}_1 \bar{z}_2 = \overline{z_1 z_2} = 9 - 7i$, $|z_1| = \sqrt{13}$, $z_1 + \bar{z}_1 = 4$, etc.

Summary of Topics

This lesson intended as a reference.

Sets and notations.

Mappings. One-to-one, onto mappings. Inverses. Composition.

Summation notation.

Items from analytic geometry and calculus.

Geometric vectors. Equality, addition, scalar multiplication. Coordinates.

Complex numbers. Conjugates, absolute value, elementary properties.

Exercises

1. Let R be the set of all real numbers, I the set of all integers, and I^+ the set of all positive integers. For $i \varepsilon R$ define $A_i = \{x \varepsilon R \mid -|i| < x < |i|\}$. Find or describe A_π, A_{-2}, A_0, $A_\pi \cap A_2$, $A_\pi \cup A_2$, $A_\pi - A_2$, $\bigcap_{i \in I} A_i$, and $\bigcap_{i \in I^+} A_i$.

2. (a) Let R = the set of all real numbers, and let $S = [0, \infty)$. Define $f: R \to S$ by $f(x) = |x|$. Is f 1-1? onto? Why?
   (b) Give an example of a mapping $f: I \to I$ which is onto but not 1-1, where I = the set of all integers.

3. Verify or evaluate:
   (a) $\sum_{j=3}^{5} (4j + j^2 + 1) = 101$.
   (b) $\sum_{i=1}^{n} \frac{1}{i^2 + i}$. (Hint: Write $\frac{1}{i^2 + i}$ as the sum of two fractions.)
   (c) $\sum_{i=1}^{2} \sum_{j=1}^{3} a_{ij} x_j = \sum_{k=1}^{3} \sum_{l=1}^{2} a_{lk} x_k$.

4. Prove the distance formula for two points $P_1(x_1, y_1, z_1)$ and $P_2(x_2, y_2, z_2)$ in three-space.

5. Find the equation of the plane which contains the points $(0, 1, 1)$, $(1, 0, 1)$, $(1, 1, 3)$.

6. Show that by adding vectors in the plane by the polygon (or parallelogram) method is equivalent to the addition $(x_1, y_1) + (x_2, y_2) = (x_1 + x_2, y_1 + y_2)$.

Lessons in Linear Algebra

7. A vector one unit long is called a *unit* vector. Find a unit vector having the same direction as the vector (2, 2, 1). (Hint: First show that the length of the vector (x, y, z) is $\sqrt{x^2 + y^2 + z^2}$.)

8. Express the vector (1, 3) as a linear combination $a(1, 1) + b(1, -1)$ of the vectors (1, 1), (1, −1). Sketch.

9. Let α, β, γ represent *three* vectors in space, neither being a linear combination of the other two. Draw a figure to illustrate how any vector δ can be expressed as a linear combination $a\alpha + b\beta + c\gamma$ of the three given vectors.

10. Bob and Phil, two swimmers of equal ability, plan to race. They depart simultaneously from a point on the bank of a stream, but they take different courses. Bob swims directly across and back along a line perpendicular to the bank, while Phil swims an equal round trip, half upstream and half downstream. Who wins?

11. For complex numbers, prove:
    (a) $\overline{z_1 z_2} = \overline{z_1}\overline{z_2}$ and $|z_1 z_2| = |z_1| |z_2|$.
    (b) $\overline{\left(\dfrac{z_1}{z_2}\right)} = \dfrac{\overline{z_1}}{\overline{z_2}}$ and $\left|\dfrac{z_1}{z_2}\right| = \dfrac{|z_1|}{|z_2|}$, if $z_2 \neq 0$.
    (c) $|z_1 + z_2| \leq |z_1| + |z_2|$ and $||z_1| - |z_2|| \leq |z_1 - z_2|$.

Remarks on Exercises

1. $A_0 = \emptyset$; $A_\pi \cap A_2 = A_2 = (-2, 2)$; $A_\pi \cup A_2 = (-\pi, \pi)$; $A_\pi - A_2 = (-\pi, -2] \cup [2, \pi)$; $\bigcap\limits_{i \in I} A_i = \emptyset$; $\bigcap\limits_{i \in I^+} A_i = A_1 = (-1, 1)$.

2. (a) The mapping f is not 1-1 because two elements have the same image. For example $f(2) = f(-2)$. The mapping f is a mapping of R onto S, because each element in S is the image of an element in R.
   (b) One example: $f: I \to I$ according to
   $$f(n) = \begin{Bmatrix} n-1 & \text{if } n \geq 0 \\ n & \text{if } n < 0 \end{Bmatrix}$$

3. (b) $\sum\limits_{i=1}^{n}\left[\dfrac{1}{i} - \dfrac{1}{i+1}\right] = 1 - \dfrac{1}{n+1}$.

5. $2x + 2y - z = 1$.

6. See Figure P-9. Let $\overrightarrow{OP}_1 = (x_1, y_1)$ and $\overrightarrow{OP}_2 = (x_2, y_2)$, and let $\overrightarrow{OQ}$ be the sum of $\overrightarrow{OP}_1$ and $\overrightarrow{OP}_2$ by the parallelogram rule. Also, let M, M_1, M_2 be the projections of Q, P_1, P_2 on the x-axis. By

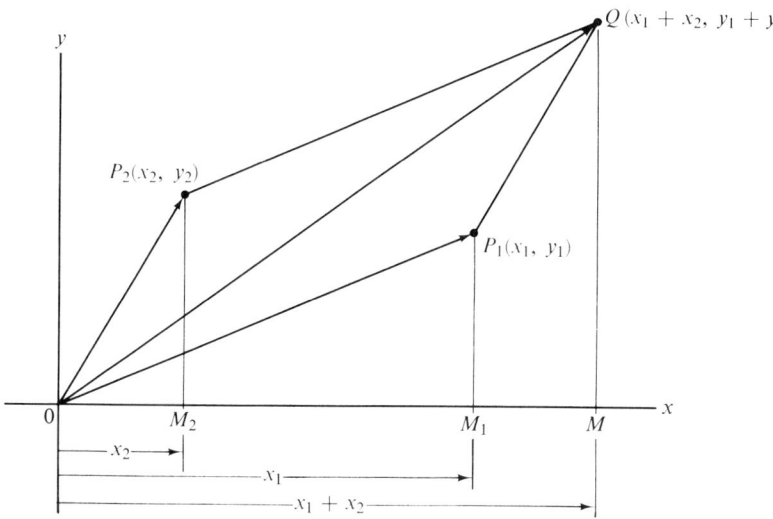

FIGURE P-9

similar triangles $OM = OM_1 + OM_2$ and hence the x-coordinate of Q is $x_1 + x_2$. The result for the y-coordinate is similar.

7. $\frac{1}{3}(2, 2, 1)$ or $\left(\frac{2}{3}, \frac{2}{3}, \frac{1}{3}\right)$.

8. $(1, 3) = 2(1, 1) - (1, -1)$.

10. Let v be the rate of each swimmer in still water, let c be the rate of the current, and let d be the width of the stream. Then compare Phil's time $[d/(v + c)] + [d/(v - c)]$ with Bob's time $(2d/\sqrt{v^2 - c^2})$. Bob wins, for $(1/\sqrt{v^2 - c^2}) < [v/(v^2 - c^2)]$, since $1 < (v/\sqrt{v^2 - c^2})$.

11. Let $z_1 = a_1 + b_1 i$, $z_2 = a_2 + b_2 i$.
    (a) $\overline{z_1 z_2} = \overline{z_1}\overline{z_2}$ by ordinary multiplication. Then $|z_1 z_2| = \sqrt{z_1 z_2 \overline{z_1 z_2}} = \sqrt{z_1 \overline{z_1}} \sqrt{z_2 \overline{z_2}} = |z_1| |z_2|$.
    (b) Similar to (a).
    (c)
    $$|z_1 + z_2|^2 = (z_1 + z_2)\overline{(z_1 + z_2)}$$
    $$= (z_1 + z_2)(\overline{z_1} + \overline{z_2})$$
    $$= z_1\overline{z_1} + z_1\overline{z_2} + z_2\overline{z_1} + z_2\overline{z_2}$$
    $$= |z_1|^2 + z_1\overline{z_2} + \overline{z_1}\,\overline{\overline{z_2}} + |z_2|^2$$
    $$= |z_1|^2 + 2\mathrm{Re}(z_1\overline{z_2}) + |z_2|^2$$
    $$\leq |z_1|^2 + 2|z_1| |z_2| + |z_2|^2$$
    $$\leq (|z_1| + |z_2|)^2.$$
    $$|z_1 + z_2| \leq |z_1| + |z_2|.$$

To get the second inequality, we apply the first to the pair $z_1 - z_2$, z_2:

$$|(z_1 - z_2) + z_2| \leq |z_1 - z_2| + |z_2|$$
$$|z_1| \leq |z_1 - z_2| + |z_2|$$
$$|z_1| - |z_2| \leq |z_1 - z_2|$$

Similarly

$$|z_2| - |z_1| \leq |z_1 - z_2|$$

Hence

$$||z_1| - |z_2|| \leq |z_1 - z_2|$$

DEFINITION AND EXAMPLES OF LINEAR SPACES (VECTOR SPACES)

In this first lesson we define the important system which we study throughout the entire book. Examples of linear spaces (sometimes called vector spaces) are provided and examined. These particular examples are used often in mathematics and we will refer to them repeatedly in the lessons which follow.

The definition of a linear space over a field is rather lengthy but important. The student should refer to it as often as he needs to so that he can recognize quickly when a given system is or is not a linear space. It is assumed that the student is familiar with fields, but as a practical matter, familiarity with the complex number field and the real number field is all that is required here. Feel free to use the rules of elementary algebra as you deal with real and complex numbers.

1-1 Definition

Let V be a set of elements called "vectors" and let F be a field of elements called "scalars." V is called a *linear space* (or *vector space*) over F if the following axioms hold for all vectors α, β, γ and all scalars a, b.

A1. $\alpha + \beta$ is a unique vector of V.
A2. $a\alpha$ is a unique vector of V.
A3. $(\alpha + \beta) + \gamma = \alpha + (\beta + \gamma)$.
A4. $\alpha + \beta = \beta + \alpha$.
A5. There is a vector 0 such that $\alpha + 0 = \alpha$.
A6. For each $\alpha \, \varepsilon \, V$ (α in V) there is a unique $-\alpha \, \varepsilon \, V$ such that $\alpha + (-\alpha) = 0$.
A7. $a(\alpha + \beta) = a\alpha + a\beta$.
A8. $(a + b)\alpha = a\alpha + b\alpha$.
A9. $(ab)\alpha = a(b\alpha)$.
A10. $1\alpha = \alpha$.

15

Lessons in Linear Algebra

The student with experience in vectors in elementary physics or geometry will observe that the vectors he studied earlier indeed satisfy the ten axioms above. This geometric origin of linear spaces should not be obscured. In fact, we will continue to borrow words like length, orthogonal, projection, and dimension from the area of geometry. Our use of these words will furthermore be analogous to their use in elementary geometry.

But we must avoid thinking that the geometric vectors familiar to us form the only vector spaces. *A vector space over a field is any system satisfying Definition 1-1.*

In order to determine if a given set V is a linear space over the field F, you should do the following:

(1) Be sure that you have clearly in mind what is the set V of vectors and what is the field F of scalars. For us, F will be either the real number field or the complex number field.

(2) Understand how to add vectors ($\alpha + \beta$) and how to take the scalar product of a vector ($a\alpha$).

(3) Check to see that all ten axioms A1, ..., A10 are satisfied. *All* must be satisfied if V is to be a linear space.

In practice one can often notice that a given structure is not a vector space by finding quickly one axiom which is not fulfilled.

Example 1. Let $V = \{(x, y) \mid x, y \in R\}$, the set of all ordered pairs of real numbers. For two vectors $\alpha = (x_1, x_2)$ and $\beta = (y_1, y_2)$, define $\alpha + \beta = (x_1 + y_1, x_2 + y_2)$ and $a\alpha = (ax_1, ax_2)$. Then V is a vector space over R, the field of real numbers. To be convinced of this, you should check points (1), (2), and (3) above. Observe that V is closed under the two operations because the sum of any two vectors (ordered pairs) is a vector and a scalar multiple of any vector is a vector. Furthermore, all of the remaining properties A1–A10 hold. Let us check two of these in detail:

A3. $(\alpha + \beta) + \gamma = \alpha + (\beta + \gamma)$, for all $\alpha, \beta, \gamma, \in V$.

Proof: Let $\alpha = (x_1, x_2)$, $\beta = (y_1, y_2)$, $\gamma = (z_1, z_2)$. Then

$$(\alpha + \beta) + \gamma = [(x_1, x_2) + (y_1, y_2)] + (z_1, z_2)$$
$$= (x_1 + y_1, x_2 + y_2) + (z_1, z_2) \quad \text{(Definition of vector addition)}$$

Definition and Examples of Linear Spaces (Vector Spaces)

$$\begin{aligned}
&= (x_1 + y_1 + z_1, x_2 + y_2 + z_2) \quad &\text{(Definition} \\
&= (x_1, x_2) + (y_1 + z_1, y_2 + z_2) \quad &\text{of vector} \\
&= (x_1, x_2) + [(y_1, y_2) + (z_1, z_2)] \quad &\text{addition)} \\
&= \alpha + (\beta + \gamma). \quad \blacksquare
\end{aligned}$$

In the proof, it is assumed that the reader is familiar with the rules for adding real numbers.

A8. $(a + b)\alpha = a\alpha + b\alpha$, for all $a, b \, \varepsilon \, R, \, \alpha \, \varepsilon \, V$.

Proof: Let $\alpha = (x_1, x_2)$. Then

$$\begin{aligned}
(a + b)\alpha &= (a + b)(x_1, x_2) \\
&= [(a + b)x_1, (a + b)x_2] \quad &\text{(Definition of scalar multiplication)} \\
&= (ax_1 + bx_1, ax_2 + bx_2) \\
&= (ax_1, ax_2) + (bx_1, bx_2) \quad &\text{(Definition of vector addition)} \\
&= a(x_1, x_2) + b(x_1, x_2) \quad &\text{(Definition of scalar multiplication)} \\
&= a\alpha + b\alpha \quad \blacksquare
\end{aligned}$$

Again, familiarity with the field operations is assumed.

The student should check the remaining axioms, but much of the proving can be done mentally without actually writing out a proof. However, you should be able to write out the proof that each of the axioms is satisfied. Note that the "zero vector" (see A5) of the above linear space is the ordered pair $(0, 0)$ of real numbers. Even though we may denote the zero vector by 0, it is understood here to be the ordered pair $(0, 0)$ of real numbers. Also, note that if $\alpha = (x, y), -\alpha = (-x, -y)$.

Example 2. Let V be the set of ordered n-tuples of real numbers and let $F = R$. Define vector addition and scalar multiplication by

$$(x_1, x_2, \ldots, x_n) + (y_1, y_2, \ldots, y_n) = (x_1 + y_1, x_2 + y_2, \ldots, x_n + y_n)$$
$$a(x_1, x_2, \ldots, x_n) = (ax_1, ax_2, \ldots, ax_n)$$

Then V is a linear space over R. (Prove it!) We denote this space R^n, and we will use it often. In particular the space of Example 1 is R^2.

Lessons in Linear Algebra

(Similarly by C^n we mean the linear space of n-tuples of complex numbers over the complex field C, where the operations of vector addition and scalar multiplication are defined coordinate-wise as above.)

Example 3. Let P be the set of all real polynomials and let $F = R$. Then P is a linear space over R.

Example 4. Let P_n be the set consisting of 0 and all polynomials of degree $\leq n$. That is let

$$P_n = \{p(x) \mid p(x) \text{ is a real polynomial of degree } \leq n\} \cup \{0\}$$

Then P_n is a linear space over R.

Example 5. Let V be the set of all real valued functions continuous on the closed interval $[0, 1]$, and let $F = R$. Let the addition and scalar multiplication of vectors (functions in this case) be the usual ones. Specifically $f + g$ is defined by $(f + g)(x) = f(x) + g(x)$ and af by $(af)(x) = af(x)$. Then V is a vector space over R.

Proof: The closure axiom A1 holds because the sum of two continuous functions is continuous. Also, if f is continuous, so is af, and the closure axiom A2 holds. The remaining axioms are easy to check. ∎

Similarly, linear spaces are formed by the set of all differentiable functions, the set of all integrable functions, and the set of all functions, on a given interval. Such linear spaces are called "function spaces" and are useful in applied mathematics and analysis. The functions you studied in calculus form linear spaces, and we will depend on your experience in calculus and analytic geometry to help you illustrate and understand linear spaces.

You may assume that every field we use henceforth is either the real number field R or the complex number field F. The linear spaces are correspondingly tagged as "real linear spaces" or "complex linear spaces" according to whether the scalars are the reals or the complexes.

Certain useful properties, shared by all linear spaces, can be proved quickly from Definition 1-1.

Definition and Examples of Linear Spaces (Vector Spaces)

Let V be a vector space over F, and let α, β, γ be general vectors, and let a be any scalar. Then

1-2
Theorem

(1) $0 + \alpha = \alpha$.
(2) $-\alpha + \alpha = 0$.
(3) $\alpha + \beta = \alpha + \gamma \Rightarrow \beta = \gamma$. (Here $\Rightarrow$ means "implies." The statement is called a "cancellation" law for vector addition.)
(4) $0\alpha = 0$.
(5) $a0 = 0$.
(6) The zero vector 0 is unique.
(7) For any α, the negative $-\alpha$ is unique.
(8) $(-1)\alpha = -\alpha$.
(9) $(-a)\alpha = a(-\alpha) = -(a\alpha)$.

Proof: Our past experience with elementary algebra may make Theorem 1-2 appear obvious. Actually (1) and (2) follow immediately because vector addition is commutative (A4) and because of A5 and A6. Also we prove (3) and (4).

(3): $\alpha + \beta = \alpha + \gamma \Rightarrow (-\alpha) + (\alpha + \beta) = (-\alpha) + (\alpha + \gamma)$
(by the uniqueness of addition)
$\Rightarrow (-\alpha + \alpha) + \beta = (-\alpha + \alpha) + \gamma$ (by A3)
$\Rightarrow \beta = \gamma$ (by (2), (1) above)

(4): $a\alpha + 0\alpha = (a + 0)\alpha$ (A8)
$= a\alpha$
$= a\alpha + 0$ (A5)

Therefore, by the cancellation law (3), $0\alpha = 0$. ∎

The other proofs are left for the student.

Summary of Topics in Lesson 1

Definition of linear space (or vector space).
Examples of linear spaces, especially R^n, P, P_n.
Function spaces.
How to determine if a given system is a linear space.
Proofs of elementary properties of all linear spaces.

Lessons in Linear Algebra

Exercises

1. Complete the verification that Examples 1 through 5 are indeed linear spaces.

2. Complete the proof of Theorem 1-2. That is, prove parts (5) through (9).

3. Let V be the set of geometric vectors (directed line segments) in a plane starting at a given point. Illustrate how an engineer or physicist would add vectors and take scalar products of vectors. Do we get a linear space? (If you have trouble starting this, consult the instructor or someone who can explain the very simple way vector addition and scalar multiplication are defined. Do not write a large amount. Just sketch a few diagrams and think.)

4. Which of the following are vector spaces? In each case it is intended that the usual operations of vector addition and scalar multiplication prevail. For example, n-tuples are added coordinate-wise and real valued functions defined on an interval are added as in your earlier courses.
   (a) the set of ordered n-tuples of complex numbers over R, the field of real numbers
   (b) the set of ordered n-tuples of real numbers over the field of complex numbers
   (c) the set $\{(x_1, x_2, x_3, x_4) \mid \text{each } x_i \in R \text{ and } \sum_{i=1}^{4} x_i = 0\}$ over R
   (d) the set of complex numbers over the field of complex numbers
   (e) the set of solutions of the differential equation $y'' - 3y' + 2y = 0$, over the real number field
   (f) the set of real polynomials of degree 6, over R
   (g) the set $V = \{(x_1, x_2)\}$ of all solutions of the equation $2x_1 - 3x_2 = 0$, over R
   (h) the set of all solutions of $2x_1 - 3x_2 = 1$, over R
   (i) the set of all real valued functions F, defined on the interval $[1, 4]$ and satisfying the condition $F(1) = F(2) = F(3) = F(4)$.
   (Answers: All except b, f, h are linear spaces.)

2

SUBSPACES, LINEAR DEPENDENCE AND LINEAR INDEPENDENCE

It is useful to consider subsets of linear spaces, especially when the subsets themselves are linear spaces. Consider for example the set V of geometric vectors in three-space, each beginning at the origin. Let W be the subset of V which is contained in a plane through the origin. Then W itself is a linear space because W is closed under vector addition and scalar multiplication. Similarly, the set of vectors on a fixed line through the origin is a linear space. In fact, the zero vector by itself forms a vector space.

We will use symbols like $\subset, \cap, \cup$ in the accepted way. $A \subset B$ means A is a *subset* of B or equivalently it means that

$$x \varepsilon A \Rightarrow x \varepsilon B$$

$A \cap B$ is the *intersection* of sets A and B, that is, $A \cap B = \{x \mid x \varepsilon A \text{ and } x \varepsilon B\}$. The *union* of A and B is the set $A \cup B = \{x \mid x \varepsilon A \text{ or } x \varepsilon B\}$.

2-1 Definition Let S be any nonempty subset of a linear space V (understood over a field F). S is called a *subspace* of V if S is a linear space (over F) under the operations of V.

Given a nonempty subset S of V, how do we decide if S is a subspace of V? We know that all of the linear space axioms A1–A10 must be satisfied, but it turns out that we have to check only two of them, the closure axioms A1 and A2.

2-2 Theorem Let S be a nonempty subset of the linear space V. Then S is a subspace of V if and only if S is closed under addition and scalar multiplication.

Proof: One part of the proof is trivial. If S is a subspace, then by definition S is a linear space and must be closed under the two operations.

Conversely, let S be a nonempty subset of V and suppose S is closed under the two operations. We must prove that S satisfies axioms A3 through A10. A1 and A2 follow from hypothesis.

Let $\alpha \in S$. Then $(-1)\alpha$ and $(-1)\alpha + \alpha$ are both vectors in S because of the closure hypotheses. This means that A6 and A5 are met. All the other axioms are satisfied by any subset of V. The proof is complete. ∎

Example 1. Let $V = R^3$ and let $W = \{(x, y, 0) \mid x, y \in R\}$. Then W is a subspace of V, because: (1) W is a nonempty subset of V; and (2) W is closed under vector addition and scalar multiplication. To understand this observe that for general vectors $(x_1, y_1, 0)$, $(x_2, y_2, 0)$ of W, the sum $(x_1 + x_2, y_1 + y_2, 0)$ is in W, and the vector $a(x_1, y_1, 0) = (ax_1, ay_1, 0)$, a vector of W.

Example 2. Let V be the linear space consisting of all real valued functions continuous on the closed interval $[a, b]$, and let W be the set of all functions differentiable on $[a, b]$. Then W is a subspace of V. Specifically, W is a subset of V because any differentiable function is continuous. Also, W is closed under the two operations because if f and g are differentiable functions, then so are $f + g$ and af.

Example 3. Let $V = R^2$ and let $W = \{(t, t + 1) \mid t \in R\}$. W is a subset but not a subspace of V, because W is not closed under vector addition. A single counterexample settles the question: The vectors $(0, 1)$ and $(1, 2)$ are in W, but their sum $(1, 3)$ is not in W. The question could have been settled just as easily either by noting that W is not closed under scalar multiplication or that W has no zero vector.

Example 4. Let V be the function space consisting of all twice-differentiable functions, and let W be the subset of V satisfying the differential equation $y'' - 2y' - 3y = 0$. W is a subspace of V because: (1) if y_1 and y_2 satisfy the differential equation, then $(y_1 + y_2)'' - 2(y_1 + y_2)' - 3(y_1 + y_2) = (y_1'' - 2y_1' - 3y_1) + (y_2'' - 2y_2' - 3y_2) = 0 + 0 = 0$, and W is closed under addition; and (2) if y_1 is solution, then ay_1 is also, for $(ay_1)'' - 2(ay_1)' - 3(ay_1) = a(y_1'' - 2y_1' - 3y_1) = a(0) = 0$.

The generalization of Example 4 is that the solution set of a linear homogeneous differential equation is a linear space which is a subspace

Subspaces, Linear Dependence and Linear Independence

of a function space. It is this fact which makes the study of linear spaces a vital foundation for the efficient study of differential equations.

Example 5. Let $V = R^4$ and let W be the solution set of the system of homogeneous linear equations

$$x_1 + 3x_2 - x_3 + 5x_4 = 0$$
$$2x_1 - x_2 + 3x_3 - x_4 = 0$$

We show that W is a subspace of R^4:

(1) Let $\alpha = (x_1, x_2, x_3, x_4)$ and $\beta = (y_1, y_2, y_3, y_4)$ be any two solutions of the system, and consider $\alpha + \beta = (x_1 + y_1, x_2 + y_2, x_3 + y_3, x_4 + y_4)$. Direct substitution into the first equation gives $(x_1 + y_1) + 3(x_2 + y_2) - (x_3 + y_3) + 5(x_4 + y_4) = (x_1 + 3x_2 - x_3 + 5x_4) + (y_1 + 3y_2 - y_3 + 5y_4) = 0 + 0 = 0$. Similarly, $\alpha + \beta$ satisfies the second equation and therefore W is closed under addition.

(2) It is also easy to verify that if α is a solution of the system, then so is $a\alpha$. This means W is closed under scalar multiplication.

The generalization of Example 5 shows that the solution set of a system of homogeneous linear equations is a linear space.

Already we can see that the study of linear spaces is relevant to linear equations, linear differential equations, n-tuple spaces, and sets of familiar functions from calculus. It is worthwhile to pause a moment and note that the things we prove for all linear spaces will hold automatically for the very special spaces of solutions to linear differential equations, solutions to linear equations in n unknowns, etc. Proofs of many properties in the more general context of an abstract linear space rather than in the special context of a particular linear space are desirable for two reasons: (1) Often the proofs are easier in the more general context; and (2) a general proof does not have to be repeated for different linear spaces.

As you study these lessons, you should frequently try to interpret your findings in the areas of linear equations, linear differential equations, and of course in n-tuple spaces. The geometric interpretation of results in two-space or three-space is often invaluable in recalling general facts and in remembering the nomenclature. A solid understanding of linear spaces will pay rich dividends in your study of mathematics and how to apply it, simply because you will be emphasizing understanding rather than memorization. Do not approach the lessons in a "theory versus application" spirit. I believe the appropriate choice is general under-

Lessons in Linear Algebra

standing of principles versus memorization and rememorization (now and hereafter).

The second main topic of this lesson is linear dependence. First we discuss a few technical terms. If A is a subset (finite or infinite) of a linear space, then a *linear combination of A* is a vector $\sum_{i=1}^{k} a_i\alpha_i$ or $a_1\alpha_1 + a_2\alpha_2 + \ldots + a_k\alpha_k$, where each $\alpha_i \in A$ and each a_i is a scalar. Note that the set $\{\alpha_1, \ldots, \alpha_k\}$ is finite, although the set A may be finite or infinite. It follows that if $A \subset B$, then any linear combination of A is also a linear combination of B. If each a_i involved in $\sum_{i=1}^{k} a_i\alpha_i$ is the zero scalar, then the linear combination $\sum_{i=1}^{k} a_i\alpha_i$ is called *trivial* and is the zero vector.

To illustrate, let $A = \{(x, x, y) \mid x, y \in R\}$. Then $2(1, 1, 3) - 5(4, 4, -7) + 3(2, 2, 0)$ is a linear combination of A. The vector $(1, 0, 0)$ cannot be written as a linear combination of A. (Try it!)

The set A of vectors is said to satisfy a *nontrivial linear relation* if there is a nontrivial linear combination of A which is equal to the zero vector, that is, if there is a "nontrivial" set of scalars $\{a_1, \ldots, a_k\}$ (not all 0) and a set $\{\alpha_1, \ldots, \alpha_k\}$ of vectors from A such that $\sum_{i=1}^{k} a_i\alpha_i = 0$.

For example, the set of vectors $\alpha_1 = (1, 2)$, $\alpha_2 = (2, 5)$, $\alpha_3 = (4, 9)$ satisfies the nontrivial linear relation $2\alpha_1 + \alpha_2 - \alpha_3 = 0$. Any nonempty set of vectors satisfies a "trivial" linear relation $0\alpha_1 + 0\alpha_2 + \ldots + 0\alpha_k = 0$.

2-3 Definition

The set A of vectors is called *linearly dependent* if there is a nontrivial linear relation satisfied by A. The set A is called linearly independent if it is not linearly dependent.

For example, the set $A = \{(1, 2), (1, 1), (5, 7), (2, 4)\}$ is linearly dependent, because $2(1, 2) + 3(1, 1) - (5, 7) = (0, 0)$ is a nontrivial linear relation of A. The nontrivial linear relation does not necessarily involve all of the vectors in A; indeed, if A is infinite, no linear relation can use all the vectors of A because the general linear relation $a_1\alpha_1 + \ldots + a_k\alpha_k = 0$ is of finite length (k terms). However, if $2\alpha_1 + 3\alpha_2 - \alpha_3 = 0$ is a nontrivial linear relation, then so is $2\alpha_1 + 3\alpha_2 - \alpha_3 + 0\alpha_4 = 0$.

You should develop and practice the determination of whether or not a given set A of vectors is linearly dependent. The following pro-

Subspaces, Linear Dependence and Linear Independence

cedure is recommended: Investigate the typical linear relation $a_1\alpha_1 + \ldots + a_k\alpha_k = 0$ of vectors in A and either: (1) display or show the existence of a nontrivial linear relation of A; or (2) show that the only linear relation is the trivial one. We have linear dependence in case (1), linear independence in case (2).

Caution: Do not waste time showing that A satisfies a trivial linear relation. This is true of any nonempty set of vectors, and is quite different from showing that the *only* linear relation satisfied by A is trivial.

Example 6. Let $A = \{\alpha_1, \alpha_2, \alpha_3\}$ where $\alpha_1 = (1, 2, 1)$, $\alpha_2 = (3, 1, 4)$, $\alpha_3 = (6, 7, 7)$. To test for linear dependence, examine the general linear relation $a_1\alpha_1 + a_2\alpha_2 + a_3\alpha_3 = 0$. This gives equivalent conditions $a_1(1, 2, 1) + a_2(3, 1, 4) + a_3(6, 7, 7) = (0, 0, 0)$ or

$$a_1 + 3a_2 + 6a_3 = 0$$
$$2a_1 + a_2 + 7a_3 = 0$$
$$a_1 + 4a_2 + 7a_3 = 0$$

Solving the above system leads to

$$a_1 + 3a_2 + 6a_3 = 0$$
$$-5a_2 - 5a_3 = 0$$
$$a_2 + a_3 = 0$$

or

$$a_3 = -a_2$$
$$a_1 = 3a_2$$

Therefore $(3a_2, a_2, -a_2)$ is the general solution of the system, with a_2 arbitrary. A simple choice $(3, 1, -1)$ gives the nontrivial linear relation $3\alpha_1 + \alpha_2 - \alpha_3 = 0$, which should be checked. The set A is therefore linearly dependent.

Example 7. Let $A = \{\alpha_1, \alpha_2, \alpha_3\}$ where $\alpha_1 = (1, 2, 1)$, $\alpha_2 = (0, 1, 1)$, $\alpha_3 = (1, 3, 4)$. We start again by considering the typical linear relation $a_1\alpha_1 + a_2\alpha_2 + a_3\alpha_3 = 0$. This gives the system

$$a_1 \phantom{{}+ a_2} + a_3 = 0$$
$$2a_1 + a_2 + 3a_3 = 0$$
$$a_1 + a_2 + 4a_3 = 0$$

Solving the system above by elementary elimination or substitution gives $a_1 = a_2 = a_3 = 0$; hence, the only linear relation is the trivial one, and A is linearly independent.

Example 8. Let V be the set of functions differentiable on the interval $(-\infty, \infty)$ and let $A = \{1, \cos 2x, \cos^2 x\}$, a subset of three differentiable functions. A is linearly dependent because with $f_1(x) = 1, f_2(x) = \cos 2x$, $f_3(x) = \cos^2 x$, we have $f_1 + f_2 - 2f_3 = 0$, a nontrivial linear relation.

Example 9. Let V be the set of functions defined on $(-\infty, \infty)$ and let $A = \{\sin x, \cos x, \sin 2x\}$. Consider the typical linear relation $a \sin x + b \cos x + c \sin 2x = 0$ (for all x). To require the relation to hold *for all* x in $(-\infty, \infty)$ is very strong. It must hold for $x = 0$. It must hold for $x = (\pi/2)$. It must hold for $x = (\pi/4)$. These three conditions alone give

$(x = 0)$ $\qquad\qquad b \qquad = 0$

$\left(x = \dfrac{\pi}{2}\right) \qquad a \qquad\qquad = 0$

$\left(x = \dfrac{\pi}{4}\right) \qquad \dfrac{1}{\sqrt{2}}a + \dfrac{1}{\sqrt{2}}b + c = 0$

and hence $a = b = c = 0$. Since the linear relation must be trivial the set A is linearly independent.

Example 10. Let $V = P$ the space of all polynomials and let $W = \{x^i \mid i = 0, 1, 2, \ldots\}$, the set of all polynomials which happen to be powers of x. We show W is a linearly independent set. Consider the general linear combination $a_0 + a_1 x + a_2 x^2 + \ldots + a_k x^k$ involving powers of x up to x^k. By the fundamental theorem of algebra the linear combination cannot be zero for all x unless each coefficient $a_0, \ldots, a_k$ is zero, because an equation of degree k cannot have more than k roots. Therefore, any linear relation of W is trivial and W is a linearly independent set. Note that W is an infinite set.

Example 11. Let V be the set of all functions continuous on the interval $I_1 = [0, 1]$ and let $W = \{x, |x|\}$, the subset consisting of two functions $f(x) = x$ and $g(x) = |x|$. On the interval I_1 we have $f(x) - g(x) = 0$, and hence W is linearly dependent.

Subspaces, Linear Dependence and Linear Independence

If we change the domain of the functions from $I_1 = [0, 1]$ to $I_2 = [-1, 1]$, we find that the set W is now linearly independent. Consider the general linear relation $ax + b|x| = 0$. If this holds for all x in I_2, then setting $x = 1$ gives $a + b = 0$ and setting $x = -1$ gives $-a + b = 0$. We are left with $a = b = 0$ and the only linear relation is trivial. So you see a set of functions may be linearly dependent on one interval, linearly independent on the other.

Finally, we observe that linear dependence is, strictly speaking, a property of an "indexed" set. For example the set of three vectors $\{\alpha, \alpha, \beta\}$ where the first two are equal is linearly dependent, while the set $\{\alpha, \beta\}$ could be linearly independent. We will distinguish between the two sets.

We will continue our study of subspaces and linear dependence in the next lesson.

Summary of Topics in Lesson 2

Definition and examples of subspaces.

How to determine whether or not a given subset is a subspace.

The listing of several areas, including linear equations and linear differential equations, unified under linear algebra.

Linear dependence and independence defined and illustrated.

Linear combinations, linear relations (trivial, nontrivial).

How to determine whether a given set is linearly dependent or linearly independent. A procedure with examples.

The linear dependence (or independence) of a subset of a function space must take into account the domain of definition of the functions.

Exercises

1. Which of the following are subspaces of R^3? Why?
    (a) $\{(x, 0, z) \mid x, z \varepsilon R\}$
    (b) $\{(x, y, 0) \mid x, y \varepsilon R \text{ and } x + y = 0\}$
    (c) $\{(x, y, 1) \mid x, y \varepsilon R\}$
    (d) $\{(x, y, x + y) \mid x, y \varepsilon R\}$
    (e) $\{(y - z, z - x, x - y) \mid x, y, z \varepsilon R\}$
    (f) $\{(x, y, z) \mid x, y, z \text{ are nonnegative}\}$
    (g) $\{(x, y, z) \mid x, y, z \text{ are integers}\}$

(h) the set consisting of just the zero vector (0, 0, 0)
(i) R^3
(j) the set of all linear combinations of (1, 1, 1) and (2, 1, 1).

2. Which of the following are subspaces of P, space of all polynomials? Why?
   (a) $\{p(x) \mid p(x) \text{ is a polynomial of degree } 5\}$
   (b) $\{p(x) \mid p(x) \text{ is a polynomial of degree } \leq 5 \text{ or } p(x) = 0\}$
   (c) $\{p(x) \mid p(x) \text{ is a polynomial divisible by } x + 1\}$
   (d) $\{p(x) \mid p(x) \text{ is a polynomial not divisible by } x + 1\}$
   (e) $\{p(x) \mid p(x) = 0 \text{ has two equal roots}\}$

3. Which of the following are subspaces of V, the space of all functions continuous on $(-\infty, \infty)$? Why?
   (a) $W = \{f \mid f' + f = 0\}$, the set of all functions f satisfying the condition $f'(x) + f(x) = 0$ for all x in $(-\infty, \infty)$
   (b) $W = \{f \mid f'(x) + f(x) = e^x \text{ for all } x\}$
   (c) $W = \{f \mid f'(x) + f(x) = 0 \text{ and } f(0) = 1\}$
   (d) P, the set of all polynomial functions of x such as $f(x) = a_0 + a_1 x + \ldots a_k x^k$
   (e) the set of all functions differentiable on $(-\infty, \infty)$
   (f) the set of all functions of the form $f(x) = a \sin x + b \cos x$, where $a, b \varepsilon R$, that is, the set of all linear combinations of $\sin x, \cos x$
   (g) the set of all linear combinations of e^x, e^{-x}

4. Let $V = R^2$. Classify each of the following subsets as linearly dependent or linearly independent:
   (a) $\{(1, 2), (3, -1), (2, 5)\}$
   (b) $\{(1, 2), (3, -1)\}$
   (c) $\{(-3, 4)\}$
   (d) $\{(-3, 4), (0, 0)\}$

5. Which of the following sets is linearly dependent in R^4?
   (a) $\{(2, 1, 4, 1), (3, -6, 2, 1), (0, 15, 8, 1)\}$
   (b) $\{(2, 1, 4, 1), (3, -6, 2, 1), (0, 15, 8, 2)\}$
   (c) $\{(1, 2, 3, 4), (7, 3, 2, 4), (0, 0, 0, 0)\}$
   (d) $\{(1, 2, 3, 4), (2, 4, 6, 8), (1, 0, 0, 0)\}$
   (e) $\{(1, 0, 0, 0), (0, 1, 0, 0), (0, 0, 1, 0), (0, 0, 0, 1)\}$

6. (a) Show that the set consisting of just the zero vector is a subspace.
   (b) Show that any set containing the zero vector is linearly dependent.

Subspaces, Linear Dependence and Linear Independence

7. Let S and T be two subspaces of V.
   (a) Show that $S \cap T$ is a subspace.
   (b) Show that $S \cup T$ is not necessarily a subspace. (Give a counter-example)

8. Explain why the empty set $\emptyset$ is taken to be linearly independent.

9. (a) Prove that if $A \subset B$ and A is linearly dependent, then B is linearly dependent.
   (b) Prove that any subset of a linearly independent set is linearly independent.

10. Given that the set $\{\alpha_1, \alpha_2, \alpha_3\}$ of three vectors is linearly independent, prove that $(\alpha_1 + \alpha_2, \alpha_2, \alpha_3)$ is linearly independent.

11. Show that a polynomial of degree 3 and its first three derivatives form a linearly independent set. Is the same true for a polynomial of degree n and its first n derivatives?

12. Determine whether or not the set $\{\cos x, \cos(x + a), \cos(x - a)\}$ of functions of x is dependent, given that a is constant.

Remarks on Exercises

1. c, f, g are not subspaces. The others are subspaces.
2. b, c are subspaces.
3. b, c are not subspaces.
4. a, d are linearly dependent sets.
5. a, c, d are linearly dependent sets.
8. $\emptyset$ does not satisfy a nontrivial linear relation.
9. (a) A nontrivial linear relation of A is a nontrivial linear relation of B.
   (b) Contrapositive of (a), logically equivalent.
11. Yes.
12. The set is linearly dependent. If $a = n\pi$, the result is easy, because then two of the functions are equal. If $a \neq n\pi$, the proof is not difficult.

3

MORE ON SUBSPACES AND LINEAR DEPENDENCE — BASES, DIMENSION

We continue our study of subspaces and linear dependence. The very important concepts of basis and dimension will be defined and illustrated. Your geometric intuition may help you here, but you should check your intuition by referring to the definitions given. For example, the dimension of R^3 is 3; however, the dimension of the subspace $\{(x, y, x + y)\}$ of R^3 is shown to be 2.

In several of the exercises of Lesson 2 we noted that the set of all linear combinations of a given subset A is a subspace. This is true generally because the sum of two linear combinations of A is a linear combination of A, and a scalar multiple of a linear combination is a linear combination.

3-1 Definition

Let A be a subset of the linear space V. By $L(A)$ we mean the set of all linear combinations of A. We call $L(A)$ the "space spanned by A" or "the space generated by A." We also say "A spans (or generates) $L(A)$."

As we have already noted $L(A)$ is a subspace of V. As a convention, for the empty set $\emptyset$ take $L(\emptyset) = \{0\}$, the subspace consisting of just the zero vector.

Example 1. Let $A = \{(1, 0, 0), (0, 1, 0)\}$, a subset of R^3. Then $L(A) = \{x(1, 0, 0) + y(0, 1, 0) \mid x, y \in R\} = \{(x, y, 0) \mid x, y \in R\}$. In a geometrical interpretation, $L(A)$ would be the xy-plane in solid analytic geometry.

Example 2. Let $A = \{\sin x, \cos x\}$. Then $L(A) = \{a \sin x + b \cos x \mid a, b \in R\}$ is a subspace of functions. The function space $L(A)$ is the complete solution set to the differential equation $y'' + y = 0$.

More on Subspaces and Linear Dependence — Bases, Dimension

Example 3. Let $A = \{(x, x) \mid x \text{ a positive integer}\} = \{(1, 1), (2, 2), \ldots\}$, a subset of R^2. Then $L(A) = \{(x, x) \mid x \varepsilon R\}$ or the set of all real multiples of $(1, 1)$, including vectors like $(-1/2, -1/2)$, (π, π), not contained in A.

The following theorem lists some simple and useful facts about $L(A)$, the space spanned by A.

3-2 Theorem

Let A be a subset of the linear space V. Then

(a) $A \subset L(A)$.
(b) $A \subset B \Rightarrow L(A) \subset L(B)$.
(c) if A is a subspace, then $L(A) = A$.
(d) $L(L(A)) = L(A)$.
(e) if $A \subset L(B)$ and $B \subset L(C)$, then $L(A) \subset L(C)$.
(f) if α is a linear combination of $A - \{\alpha\}$ (the other vectors of A), then $L(A - \{\alpha\}) = L(A)$.

Proof: The proofs are easy. (a) Each vector in A is a linear combination of A. (b) If A is a subset of B then each linear combination of A is a linear combination of B. (c) $A \subset L(A)$ from part (a). Also $L(A) \subset A$ because if A is a subspace, it must be closed under linear combinations. Part (d) is an application of part (c), with A replaced by $L(A)$. (e) $A \subset L(B) \Rightarrow L(A) \subset L(L(B)) = L(B)$ by a, d. Similarly, $B \subset L(C) \Rightarrow L(B) \subset L(C)$. Hence, $L(A) \subset L(C)$. (f) Since $A - \{\alpha\} \subset A$, part (a) gives $L(A - \{\alpha\}) \subset L(A)$. To get inclusion the other way, we observe that $A \subset L(A - \{\alpha\})$ by hypothesis. Then $L(A) \subset L(L(A - \{\alpha\})) = L(A - \{\alpha\})$ by parts (a), (d). ∎

3-3 Definition

A *basis* of the linear space V is a linearly independent set which spans V.

For the set A to be a basis of V, the following *two* conditions must be met:

(1) A spans V. That is, $L(A) = V$.
(2) A is linearly independent.

It is important to understand that one of these conditions may hold, while the other fails. In order to show that a set is a basis you must prove (1) and (2) separately.

Lessons in Linear Algebra

Example 4. We show that $A = \{(1, 1), (1, -1)\}$ is a basis of R^2. (1) In order to show that the subset A spans R^2 we must show that the general vector (x, y) of R^2 is a linear combination of A. You can easily verify that $(x, y) = [(x + y)/2](1, 1) + [(x - y)/2](1, -1)$, a linear combination of A. (Getting the correct pair of scalars $(x + y)/2$ and $(x - y)/2$ comes from solving the vector equation $(x, y) = a(1, 1) + b(1, -1)$ for a and b. That is, we solve the system $a + b = x$, $a - b = y$, for a, b in terms of x, y.)

(2) We must also show that A is linearly independent. Consider the typical linear relation $c_1(1, 1) + c_2(1, -1) = (0, 0)$. This is equivalent to $c_1 + c_2 = 0$, $c_1 - c_2 = 0$ or to $c_1 = c_2 = 0$. The only linear relation of A is trivial and, therefore, A is linearly independent.

Example 5. Is the set $A = \{(1, 0, 0), (0, 1, 0)\}$ a basis of R^3? The set A is clearly linearly independent, because $a(1, 0, 0) + b(0, 1, 0) = (0, 0, 0)$ implies $a = b = 0$. But not every vector in R^3 is a linear combination of A. For example $(0, 0, 1) = a(1, 0, 0) + b(0, 1, 0)$ cannot be satisfied because $(0, 0, 1) \neq (a, b, 0)$. A is not a basis of R^3, even though A is linearly independent. (It is true however that A is a basis of the subspace $L(A)$. In fact any linearly independent set A is a basis of $L(A)$.)

Example 6. (1) The set $A = \{(1, 0), (1, 1), (1, -1)\}$ spans R^2, because the general vector of R^2 is a linear combination of A. In fact this can be done in more than one way.

$$(x, y) = (x - y)(1, 0) + y(1, 1)$$

or

$$(x, y) = \frac{x + y}{2}(1, 1) + \frac{x - y}{2}(1, -1)$$

or

$$(x, y) = (x - 3y)(1, 0) + 2y(1, 1) + y(1, -1)$$

(2) The set A is linearly dependent, because the linear relation $a(1, 0) + b(1, 1) + c(1, -1) = (0, 0)$ can be satisfied nontrivially by (for example) $a = -2$, $b = c = 1$.

Therefore, A is not a basis of R^2, although A spans R^2. An additional observation is in order. If the linear combination giving (x, y) above had been unique, then the spanning set A would have been a basis, and conversely. This we will prove shortly.

More on Subspaces and Linear Dependence — Bases, Dimension

Example 7. The set $A = \{1, x, x^2, \ldots, x^n, \ldots\}$ is a basis for P, the space of all polynomials. In Lesson 2 (Example 10) we showed A to be linearly independent. Also, any polynomial is in $L(A)$.

Example 8. Consider the linear homogeneous differential equation $ay''' + by'' + cy' + dy = 0$. It is easy to show that e^{rx} is a solution if and only if r is a root of the polynomial equation $ax^3 + bx^2 + cx + d = 0$. Substituting e^{rx} for y gives $(ar^3 + br^2 + cr + d)e^{rx} = 0$, which holds if and only if $ar^3 + br^2 + cr + d = 0$. If there are n distinct real roots to the "auxiliary" equation $ax^3 + bx^2 + cx + d = 0$, then we get three solutions $A = \{e^{r_1 x}, e^{r_2 x}, e^{r_3 x}\}$. The entire solution set of the differential equation is $L(A)$, a result which is extended and used often in differential equations. You will be asked to show that the set A is linearly independent as one of the exercises, but we will not show that every solution must take the form $c_1 e^{r_1 x} + c_2 e^{r_2 x} + c_3 e^{r_3 x}$.

Example 9. We illustrate how to find a basis of the subspace $W = \{(x, y, x + y)\}$ of R^3. First we write W as $x(1, 0, 1) + y(0, 1, 1)$, which tells us that each vector of W is a linear combination of the two-element set $A = \{(1, 0, 1), (0, 1, 1)\}$. We then have that A spans W or $L(A) = W$, and we need only to check that A is linearly independent. The linear relation $x(1, 0, 1) + y(0, 1, 1) = (0, 0, 0)$ is satisfied only when $(x, y, x + y) = (0, 0, 0)$ or when $x = y = 0$. Therefore A is a linearly independent set which spans W, that is, A is a basis of W.

As you may expect, a linear space has more than one basis. Look at R^2. $A = \{(1, 0), (0, 1)\}$ is a basis. $B = \{(1, 1), (1, -1)\}$ is a basis. We call A the "standard" basis for R^2, and we extend this notion to R^n. It is very easy to express the typical vector (x, y) as a linear combination of A, $(x, y) = x(1, 0) + y(0, 1)$.

3-4 Definition

The *dimension* of a linear space V, denoted dim V, is the number of elements in a basis of V. If V has a finite basis, then V is called a *finite-dimensional* linear space. Otherwise V is called *infinite-dimensional*. Most of our work will be with finite-dimensional spaces.

At first glance we should wonder whether or not a given space could have two bases of different orders. This, of course, would make the definition ambiguous and unacceptable. Fortunately, as we shall see

Lessons in Linear Algebra

in the next lesson, any two bases have the same number of elements. For the time being, we accept this without proof.

In order to find dim V, it suffices simply to count the number of elements in any basis. Example 4 shows dim $(R^2) = 2$. Similarly dim $(R^n) = n$. Example 7 shows P is infinite-dimensional. By Example 9 dim $\{(x, y, x + y)\} = 2$, because the basis $\{(1, 0, 1), (0, 1, 1)\}$ has two elements.

3-5 Theorem

Let A be a set which spans the linear space V. Then A is a basis of V if and only if the representation of each vector β of V in terms of A is unique.

Proof: We are given $L(A) = V$. For the first part of the proof, let A be a basis of V, and let β be a typical vector of V. Suppose β is equal to two linear combinations of A, one linear combination of the subset A_1, and another linear combination of the subset A_2. Then $\beta = a_1\alpha_1 + \ldots + a_n\alpha_n = b_1\alpha_1 + \ldots + b_n\alpha_n$, where $A_1 \cup A_2 = \{\alpha_1, \ldots, \alpha_n\}$ and some of the scalars a_i, b_i may be zero. Then we have $(a_1 - b_1)\alpha_1 + (a_2 - b_2)\alpha_2 + \ldots + (a_n - b_n)\alpha_n = 0$. Since the basis A is linearly independent, the triviality of the preceding relation gives $a_i - b_i = 0$ or $a_i = b_i$ for each i. Therefore, the two representations are the same.

Conversely, suppose each vector β of V has a unique representation $\beta = a_1\alpha_1 + \ldots + a_n\alpha_n$ in terms of the spanning set A. Then A must be linearly independent (and hence a basis), because otherwise there would be a nontrivial linear relation $b_1\alpha_1 + \ldots + b_n\alpha_n = 0$. This would contradict the uniqueness of the representation of the zero vector. ∎

The following theorem has a usefulness all out of proportion to its simplicity.

3-6 Theorem

The set A is linearly dependent if and only if one vector of A is a linear combination of the others.

Proof: If A is linearly dependent, then a nontrivial linear relation $a_1\alpha_1 + \ldots + a_n\alpha_n = 0$ with $\alpha_k \neq 0$ gives

$$-a_k\alpha_k = \sum_{\substack{i=1 \\ i \neq k}}^{n} a_i\alpha_i$$

More on Subspaces and Linear Dependence — Bases, Dimension

and then multiplying by $-1/a_k$ gives

$$\alpha_k = \sum -\frac{a_i}{a_k}\alpha_i$$

a linear combination of $A - \{\alpha_k\}$.

Conversely, if some vector, say α_k, is a linear combination of $A - \{\alpha_k\}$, then $\alpha_k = \sum a_i\alpha_i$ leads to the nontrivial linear relation $-\alpha_k + \sum a_i\alpha_i = 0$, and the set A is linearly dependent. ∎

We conclude this lesson by considering the intersection and union of subspaces. The intersection of subspaces is a subspace, but the union of subspaces is not in general a subspace.

3-7 Theorem The intersection of any collection of subspaces of V is a subspace.

Proof: Let $\{A_i\}$ where i is in some indexing set I be a collection of subspaces of V, and consider $B = \bigcap_{i \in I} A_i$, the intersection of all the subspaces A_i. If $\alpha, \beta \in B$, then α, β belong to each A_i. Since each A_i is a subspace, then $\alpha + \beta$ and $a\alpha$ are both in each A_i. This means $\alpha + \beta$ and $a\alpha$ both belong to B; it follows that B is a subspace. ∎

The union of two subspaces is not, in general, a subspace, as the counterexample (Example 10) shows. However, the space spanned by the union of subsets is given a special name.

3-8 Definition Let S and T be two subspaces of the vector space V. We define the *sum of S and T*, denoted $S + T$, by

$$S + T = \{\alpha \mid \alpha = \sigma + \tau, \text{ where } \sigma \in S, \tau \in T\}$$

You may think of $S + T$ as the set of vectors obtained by adding vectors in S to vectors in T.

Example 10. Consider $S = \{(x, y, 0) \mid x, y \in R\}$ and $T = \{(0, 0, z) \mid z \in R\}$, two subspaces of R^3. The union $S \cup T$ is not a subspace because, for one thing, the sum of $\sigma = (1, 1, 0)$ and $\tau = (0, 0, 1)$ is not in $S \cup T$. The sum $S + T$ is a subspace; in fact, $S + T = R^3$, because the general vector (x, y, z) of R^3 can be written $(x, y, 0) + (0, 0, z)$, a vector of $S + T$. The general result is given as a theorem.

Lessons in Linear Algebra

3-9 Theorem

Let S and T be subspaces of the vector space V. Then $S + T$ is the space spanned by $S \cup T$. That is, $L(S \cup T) = S + T$.

Proof: First we have $S + T \subset L(S \cup T)$, because the typical vector $\sigma + \tau$ is a linear combination of $S \cup T$. Conversely, we must show $L(S \cup T) \subset S + T$. Let $\alpha \in L(S \cup T)$, with $\alpha = \sum a_i \sigma_i + \sum b_i \tau_i$, where each $\sigma_i \in S$ and each $\tau_i \in T$. Then α is the sum of two vectors $\sum a_i \sigma_i$ and $\sum b_i \tau_i$, the first in S (because S is a subspace) and the second in T. Hence, $\alpha \in S + T$, and $L(S \cup T) \subset S + T$.

Since we have containment both ways, we know that $L(S \cup T) = S + T$. ∎

Summary of Topics in Lesson 3

The set $L(A)$ of linear combinations of A. Spanning sets.

Basis of a linear space. How to determine if a given set is a basis. How to find a basis of a given space.

Dimension of a linear space V, denoted dim V. How to find dim V.

The unique representation of a vector in terms of a given basis.

The intersection and union of subspaces.

The sum of two subspaces.

Exercises

1. Show that $A = \{(1, 0, 0, 0), (0, 1, 0, 0), (0, 0, 1, 0), (0, 0, 0, 1)\}$ is a basis of R^4. Extend the idea and describe the "standard basis" of R^n.

2. (a) Is the set $\alpha_1 = (2, 4)$, $\alpha_2 = (-1, -2)$ a basis of R^2? Describe the space $L(A)$ spanned by $A = \{\alpha_1, \alpha_2\}$.
   (b) Is the set $\alpha_1 = (3, 1)$, $\alpha_2 = (-1, 2)$ a basis of R^2? Express $(-4, 5)$ as a linear combination of α_1, α_2.
   (c) What is the geometric relation between the two bases $A = \{(1, 0), (0, 1)\}$ and $B = \{(1/\sqrt{2}, 1/\sqrt{2}), (-1/\sqrt{2}, 1/\sqrt{2})\}$? Express each vector of B in terms of A, and vice versa.

3. (a) Is the set $\alpha_1 = (1, 1, 0)$, $\alpha_2 = (-1, 4, 2)$, $\alpha_3 = (0, 0, 2)$ a basis of R^3? Express $\beta = (-5, 10, -2)$ as a linear combination of $\{\alpha_1, \alpha_2, \alpha_3\}$. Is this representation unique?
   (b) Is the set $\alpha_1 = (1, 1, -1)$, $\alpha_2 = (2, 0, 3)$, $\alpha_3 = (0, 2, -5)$ a basis of R^3? Express $\beta = (4, 2, 1)$ as a linear combination of

More on Subspaces and Linear Dependence — Bases, Dimension

$\{\alpha_1, \alpha_2, \alpha_3\}$. Is this representation unique? Try the same for the vector $\gamma = (0, 0, 1)$.

4. Let V be the subspace $\{(x, y, x + y, x - y)\}$ of R^4. Find a basis of V and the dimension of V.

5. Let $V = \{(x_1, x_2, x_3, x_4, x_5) \mid x_1 = x_3 = x_5\}$, a subspace of R^5. Find a basis of V and dim V.

6. Let $V = \{(x_1, x_2, \ldots, x_n) \mid \sum_{i=1}^{n} x_i = 0\}$, a subspace of R^n. Find a basis of V such that for each vector $(x_1, x_2, \ldots, x_n)$ in the basis, $\sum_{i=1}^{n} x_i^2 = 1$. Find dim V.

7. The set A of vectors is called a maximal *linearly independent set* if: (1) A is linearly independent and (2) any set B which *properly* contains A is linearly dependent. Show that a maximal linearly independent set of the linear space V is a basis of V.

8. Show that a minimal spanning set of V is a basis of V. (A is a minimal spanning set of V if A spans V, and no proper subset of A spans V.)

9. The solution set S of the system

$$x + y + z - 4w = 0$$
$$2x + 3y + z + w = 0$$

is a subspace of R^4. Find a basis of S and the dimension of S.

10. Let $S = \{(x, y, x - y)\}$ and $T = \{(x, -x, y)\}$ be subspaces of R^3.
    (a) Find a basis of S.
    (b) Find a basis of T.
    (c) Find a basis of $S \cap T$.
    (d) Find a basis of $S + T$.
    (e) Compare the dimensions of the above four subspaces. We will show in the next lesson that dim S + dim T = dim $(S \cap T)$ + dim $(S + T)$.

11. Let P be the space of polynomials and let S and T be subspaces of P. Specifically, let S be the set of polynomials having $x + 1$ as a factor, and let T be the set of polynomials having $x - 1$ as a factor. Describe $S \cap T$ and $S + T$. Find a basis for each space and the dimension of each space.

12. Find a basis for the solution space S of the differential equation $y'' - 7y' + 12y = 0$, given that dim $S = 2$.

13. Prove that if $\{\alpha_1, \alpha_2\}$ is a basis of V, then $\{\alpha_1 + \alpha_2, \alpha_1 - \alpha_2\}$ is a basis.

14. Let V be the vector space consisting of all geometric vectors in three-space, and let S, T, and U be subsets: (1) S is the set of all vectors on the x-axis; (2) T is the set of vectors on the plane $z = 0$; and (3) U is the set of all vectors on a fixed line through the origin but outside the plane $z = 0$.
 Describe geometrically each of the following subspaces: $V \cap V$, $S \cap T$, $S \cup T$, $S + T$, $T \cap U$, $U + S$, $T + T$, $T + U$. Is it true in this example that $S \cap (T + U) = (S \cap T) + (S \cap U)$? Is $S \cap (T + U) = T + (S \cap U)$? Is $S \cap (T + U) \subset T + (S \cap U)$?

15. Let S, T, U be subspaces of V. Prove:
 (a) $S \subset S + T$.
 (b) if $S \subset U$ and $T \subset U$, then $S + T \subset U$,
 (c) $S \cap T \subset S$.
 (d) $S \cap (S + T) = S$.
 (e) $S + (S \cap T) = S$.
 (f) $(S \cap T) + (S \cap U) \subset S \cap (T + U)$.

Remarks on Exercises

2. (a) No. The set is linearly dependent since $\alpha_1 = -2\alpha_2$. $L(A)$ is the space of all multiples of α_1.
 (b) Yes. The set is linearly independent. Also the set spans R^2, because $(x, y) = [(2x + y)/7](3, 1) + [(-x + 3y)/7](-1, 2)$. $(-4, 5) = -(3/7)(3, 1) + (19/7)(-1, 2)$.
 (c) A rotation of 45° takes A into B.

3. (a) Yes, the set is a basis because it is linearly independent and also the general vector
$$(x, y, z) = \frac{4x + y}{5}(1, 1, 0) + \frac{-x + y}{5}(-1, 4, 2)$$
$$+ \frac{2x - 2y + 5z}{10}(0, 0, 2)$$
And
$(-5, 10, -2) = -2(1, 1, 0) + 3(-1, 4, 2) - 4(0, 0, 2)$
uniquely.

More on Subspaces and Linear Dependence — Bases, Dimension

(b) No, the set is linearly dependent. The expression of β in terms of $\{\alpha_i\}$ is not unique. For example, $\beta = 2\alpha_1 + \alpha_2 = 2\alpha_2 + \alpha_3 = -2\alpha_1 + 3\alpha_2 + 2\alpha_3$, to list a few.
The vector $(0, 0, 1)$ is not a linear combination of $\alpha_1, \alpha_2, \alpha_3$.

4. A basis of V is $\{(1, 0, 1, 1), (0, 1, 1, -1)\}$. $\dim V = 2$.

5. A basis of V is $\{(1, 0, 1, 0, 1), (0, 1, 0, 0, 0), (0, 0, 0, 1, 0)\}$. $\dim V = 3$.

6. A basis of V is the set of $n-1$ vectors $\{(1, 0, 0, \ldots, 0, -1), (0, 1, 0, \ldots, 0, -1), \ldots, (0, 0, \ldots, 0, 1, -1)\}$, so $\dim V = n-1$, To make $x_i^2 = 1$, multiply each vector in the basis by $1/\sqrt{2}$.

7. We show that if A is a linearly independent set which is not a basis, then A cannot be maximal. If A is not a basis, then there is a vector $\beta \notin L(A)$. Then $A \cup \{\beta\}$ is linearly independent which contradicts the hypothesis that A is maximal.

8. Show that a spanning set which is not a basis cannot be minimal.

9. The solution set is $S = \{(-2z + 13w, z - 9w, z, w) \mid z, w \in R\}$. A basis of S is $\{(-2, 1, 1, 0), (13, -9, 0, 1)\}$, and $\dim S = 2$.

10. (a) $\{(1, 0, 1), (0, 1, -1)\}$; (b) $\{(1, -1, 0), (0, 0, 1)\}$; (c) $\{(1, -1, 2)\}$; (d) $\{(1, -1, 2), (1, 0, 1), (0, 0, 1)\}$. (There are other choices, since $S + T = R^3$). (e) $\dim S = 2$; $\dim T = 2$; $\dim (S \cap T) = 1$; $\dim (S + T) = 3$.

11. A basis of S is the infinite set of polynomials $\{x - 1, (x - 1)x, (x - 1)x^2 \ldots\}$. A basis of T is $\{x + 1, (x + 1)x, (x + 1)x^2, \ldots\}$. The space $S \cap T = \{(x - 1)(x + 1)p(x) \mid p(x) \text{ a polynomial}\}$ and has infinite basis $\{(x - 1)(x + 1)x^i \mid i = 0, 1, \ldots\}$. The space $S + T = P$, the entire space of polynomials. To see this consider the identity $p(x) = -(1/2)p(x)(x - 1) + (1/2)p(x)(x + 1)$ which shows $P = S + T$. Each of the four spaces is infinite-dimensional.

12. $\{e^{3x}, e^{4x}\}$.

14. $V \cap V = V$, $S \cap T = S$, $S \cup T = T$, $S + T = T$, $T \cap U = \{0\}$ $U + S$ is a plane containing U and S, $T + T = T$, $T + U = V$. $S \cap (T + U) = S$, so the answers to the last three questions are yes, no, yes.

15. (f) Let $\alpha \in (S \cap T) + (S \cap U)$. Then $\alpha = \beta + \gamma$ where $\beta \in S \cap T$ and $\gamma \in S \cap U$. Then β and γ are in S and hence $\alpha \in S$. We need to show α is also in $T + U$. This is true because $\alpha = \beta + \gamma$ where $\beta \in T$ and $\gamma \in U$.

4

MORE ON BASES, SPANNING SETS, AND DIMENSION

This lesson has three main purposes: (1) to show that the dimension of a linear space is well-defined; (2) to refine methods for finding certain bases of a given linear space and for determining the dimension of the space; and (3) to compare the subspaces obtained by taking the intersection and sum of two spaces.

We have already shown that a set A is linearly dependent if and only if one vector of A is a linear combination of the others (Theorem 3-6). A similar but different result follows.

4-1 Theorem

Let $A = \{\alpha_1, \alpha_2, \ldots, \alpha_n, \ldots\}$ be a sequence of nonzero vectors from a linear space V. Then A is linearly dependent if and only if one vector of A is a linear combination of the preceding vectors.

Proof: First, suppose one vector of A is a linear combination of the preceding. Then by Theorem 3-6 the set A is linearly dependent.

The other half of the proof is left for the student, with a few remarks. The hypothesis that the set A is a sequence of nonzero vectors is essential. Also, the conclusion that a vector α_k is a linear combination of the preceding is stronger than the statement that α_k is a linear combination of the others. As an illustration, the linearly dependent set $\{0, \alpha, \beta\}$ enjoys the property that one is a linear combination of the others but may not satisfy the condition that one is a linear combination of the preceding. ∎

The following theorem shows that the definition of dimension is unambiguous.

More on Bases, Spanning Sets, and Dimension

Let V have the finite basis $B = \{\beta_1, \beta_2, \ldots, \beta_n\}$. Then

4-2 Theorem

(a) any linearly independent set has at most n elements.
(b) any two bases of V have the same number of elements.
(c) any linearly independent set can be extended to a basis of V.

Proof: (a) We proceed by repeating a simple step. Let $A = \{\alpha_1, \ldots, \alpha_k\}$ be linearly independent, and consider the dependent set $\{\alpha_1, \beta_1, \ldots, \beta_n\}$ of $n + 1$ vectors, in which one of the β's is a linear combination of the preceding vectors by Theorem 4-1. We delete this β and get the n-element set

$$A_1 = \{\alpha_1, \beta_2, \ldots, \beta_n\}$$

which spans V by Theorem 3-2(f). (Relabel the β's, if necessary.)

Now we repeat the entire process. In the dependent set $\{\alpha_1, \alpha_2, \beta_2, \ldots, \beta_n\}$ of $n + 1$ elements, one β must be a linear combination of the preceding vectors by Theorem 4-1. Dropping this β gives the n-element spanning set

$$A_2 = \{\alpha_1, \alpha_2, \beta_3, \ldots, \beta_n\}$$

(Again, relabel the β's if necessary.)

After k such steps, we have the n-element spanning set

$$A_k = \{\alpha_1, \ldots, \alpha_k, \beta_{k+1}, \ldots, \beta_n\}$$

Note that $k \leq n$, because if $k > n$, then $A_n = \{\alpha_1, \ldots, \alpha_n\}$ would be a basis, contradicting the linear independence of $\{\alpha_1, \ldots, \alpha_k\}$.

This completes the proof of (a) by showing that in a space with an n-element basis, no linearly independent set (finite or infinite) can contain a subset of more than n-elements.

(b) Let C be a basis with m elements. Then $m \leq n$. Reversing the roles played by B and C gives $n \leq m$. Hence $m = n$.

(c) The constructive nature of the proof of part (a) shows how any linearly independent set can be extended to a basis. The spanning set A_k is a basis because otherwise it would be linearly dependent. But this would mean that a basis of less than n elements could be formed by dropping some of the elements of A_k, and would contradict part (b). ∎

Your understanding of the above theorem (and its proof) may be helped by an illustration.

Example 1. Let $V = R^4$ and let B be the standard basis, $\beta_1 = (1, 0, 0, 0)$, $\beta_2 = (0, 1, 0, 0)$, $\beta_3 = (0, 0, 1, 0)$, $\beta_4 = (0, 0, 0, 1)$. Let A consist of the two vectors $\alpha_1 = (1, 1, 1, 1)$, $\alpha_2 = (1, -1, 1, -1)$. We propose to extend $A = \{\alpha_1, \alpha_2\}$ to a basis by following the steps used in the proof of Theorem 4-2.

Step 1. Consider $\{\alpha_1, \beta_1, \beta_2, \beta_3, \beta_4\}$, and note that $\beta_4 = \alpha_1 - \beta_1 - \beta_2 - \beta_3$. Dropping β_4 gives the basis $A_1 = \{\alpha_1, \beta_1, \beta_2, \beta_3\}$.

Step 2: Now consider $\{\alpha_1, \alpha_2, \beta_1, \beta_2, \beta_3\}$. Since $\beta_3 = (1/2)\alpha_1 + (1/2)\alpha_2 - \beta_1$, dropping β_3 gives the required basis $A_2 = \{\alpha_1, \alpha_2, \beta_1, \beta_2\}$.

In constructing A_2 we have extended A to a basis of V by adding two elements from the basis B. The construction is not unique.

There are immediate corollaries to Theorem 4-2.

4-3 Corollary

Let V be an n-dimensional linear space. Then

(a) let A be a linearly independent set. Then A is a basis of V if and only if A has exactly n elements.

(b) let A be a set which spans V. Then A is a basis of V if and only if A has exactly n elements.

(c) any set which spans V contains a basis of V.

Proof: (a) Let A be linearly independent. If A is a basis, then A has exactly n elements by Theorem 4-2(b). Conversely, if A has exactly n elements then A must be a basis, or else A could be extended to a basis of more than n elements, contradicting Theorem 4-2(b).

(b) Let A span V. If A is a basis, then A has exactly n elements by Theorem 4-2(b). Conversely, suppose A has exactly n elements. A must be linearly independent, because otherwise A would contain a maximal linearly independent subset B with less than n elements, and B would be a basis of V because $A \subset L(B)$ and hence $L(A) \subset L(B)$. (See Theorem 3-2(b), (d), (f).)

(c) Let A span V. If A has exactly n elements, we apply part (b). If A has more than n elements (perhaps an infinite number), then we can construct a basis $\{\alpha_1, \alpha_2, \ldots, \alpha_n\}$ consisting of vectors from A. First select $\alpha_1 \neq 0$. Then choose α_2 such that $\{\alpha_1, \alpha_2\}$ is linearly independent. We can continue this to get $\{\alpha_1, \alpha_2, \ldots, \alpha_n\}$, because otherwise a smaller set would generate V. This is impossible because no basis contains fewer than n elements. ∎

The proof above shows also that any "finitely-generated" space has a finite basis. That is, if V has a finite spanning set B, then B contains a finite basis of V. It is also true that any linear space has a basis, but the

More on Bases, Spanning Sets, and Dimension

proof of this is beyond the scope of this book. (For further details, consult Nathan Jacobson, *Lectures in Abstract Algebra*, vol. 2, Van Nostrand, Princeton, 1960.)

It is useful to know that any linearly independent set can be extended to a basis, and that any spanning set contains a basis. Sometimes we use these facts without going through the tedium of actually working out the details. Suppose A is a linearly independent set in an n-dimensional space. We may say "Extend A to the basis B," with the understanding that B is a basis containing A as a subset. The fact that this can be done in more than one way and the tedious procedure of the extension may or may not concern us.

Example 2. Let V be the function space spanned by the basis $B = \{e^x, e^{-x}, e^{2x}\}$ and let $A = \{\sinh x\}$. Recall $\sinh x = (1/2)e^x - (1/2)e^{-x}$. Therefore, $A \subset V$ and A can be extended to a basis of V.

Here is one method for actually extending A to a basis. The three-element set $C = \{\sinh x, e^x, e^{2x}\}$ spans V, because each vector of B is a linear combination of C. Therefore, C is a basis by Theorem 4-3(b).

We were assured that the spanning set $\{\sinh x, e^x, e^{-x}, e^{2x}\}$ contained a basis by Corollary 4-3(c).

Example 3. Let P_4 be the space of polynomials of degree ≤ 4, and let $A = \{(x-1)x^3, (x-2)x^3, (x-3)x^3, (x-4)x^3, \ldots\}$, an infinite subset of P_4. Let us test the first four for linear dependence. Consider the linear relation $a(x-1)x^3 + b(x-2)x^3 + c(x-3)x^3 + d(x-4)x^3 = 0$ (for all x), equivalent to $(a+b+c+d)x^4 - (a+2b+3c+4d)x^3 = 0$ (for all x).

The nontrivial solution $a = 1$, $b = -2$, $c = 1$, $d = 0$ shows that the first four elements of A form a linearly dependent set. This is not surprising. A glance at A suggests that the polynomials generated by A contain only the powers x^3 and x^4.

Let $B = \{x^3, x^4\}$. Since B is linearly independent, B is a basis of $L(B)$. The set $A \subset L(B)$ because $(x-k)x^3 = x^4 - kx^3$, and hence, dim $L(A) \leq 2$. Any linearly independent two-element subset of A is a basis of $L(A)$. The set $\{(x-1)x^3, (x-2)x^3\}$ is such a basis since neither function is a scalar multiple of the other.

The set $\{(x-1)x^3, (x-2)x^4\}$ can be extended to the basis $\{(x-1)x^3, (x-2)x^4, 1, x, x^2\}$ of P_4.

Now we compare the four spaces S, T, $S \cap T$, $S + T$. Their dimensions and their bases are closely related.

Lessons in Linear Algebra

4-4 Theorem

Let S and T be subspaces of a finite-dimensional space. Then

$$\dim S + \dim T = \dim (S \cap T) + \dim (S + T)$$

Proof: Let the dimensions of S, T, $S \cap T$, $S + T$ be respectively s, t, m, n, and let $A = \{\alpha_1, \alpha_2, \ldots, \alpha_m\}$ be a basis of $S \cap T$. Extend A to the basis $\{\alpha_1, \alpha_2, \ldots, \alpha_m, \beta_1, \beta_2, \ldots, \beta_{s-m}\}$ of S. Similarly, let $\{\alpha_1, \ldots, \alpha_m, \gamma_1, \ldots, \gamma_{t-m}\}$ be a basis of T.

We show that the set $B = \{\alpha_1, \ldots, \alpha_m, \beta_1, \ldots, \beta_{s-m}, \gamma_1, \ldots, \gamma_{t-m}\}$ is a basis of $S + T$. Clearly B spans $S + T$, because B spans S and B spans T. It remains to be shown that B is linearly independent. If

$$a_1\alpha_1 + \ldots + a_m\alpha_m + b_1\beta_1 + \ldots + b_{s-m}\beta_{s-m} + c_1\gamma_1 + \ldots + c_{t-m}\gamma_{t-m} = 0$$

then the vector

$$\delta = -\sum_{i=1}^{t-m} c_i\gamma_i = \sum_{i=1}^{m} a_i\alpha_i + \sum_{i=1}^{s-m} b_i\beta_i$$

and is in T (look at the left member) and also in S (look at the right member). The unique representation of δ according to the basis $\{\alpha_1, \ldots, \alpha_m\}$ and according to the basis $\{\alpha_1, \ldots, \alpha_m, \beta_1, \ldots, \beta_{s-m}\}$ shows that each $b_i = 0$.

A similar consideration of the equation $-\sum b_i\beta_i = \sum a_i\alpha_i + \sum c_i\gamma_i$ shows that each $c_i = 0$. Then $\sum a_i\alpha_i = 0$ and each $a_i = 0$ because the set $\{\alpha_1, \ldots, \alpha_m\}$ is linearly independent.

Now that we have established that B is a basis of $S + T$, we compare dimensions. Recall $\dim S = s$, $\dim T = t$, and $\dim (S \cap T) = m$. Counting the elements in B shows that $\dim (S + T) = s + t - m$. Therefore, $\dim (S + T) = \dim S + \dim T - \dim (S \cap T)$, and the proof is complete. ∎

Theorem 4-4 was illustrated by Exercise 10 of Lesson 3. The important thing to remember is that if we know the dimensions of three of the spaces S, T, $S \cap T$, $S + T$, then we can find the dimension of the fourth by simple arithmetic. If we were interested, for example, in finding a basis of a space, it helps if we know the dimension of the space in advance.

Example 4. Let S be the subspace of R^4 spanned by $\{(1, 2, 1, -1), (2, 0, 1, 1), (3, 2, 2, 0)\}$, and let T be the space spanned by $\{(4, 4, 3, -1), (1, 0, 0, 0)\}$. For each of the spaces $S \cap T$ and $S + T$ we find a basis.

More on Bases, Spanning Sets, and Dimension

In the spanning set for S, the third vector is the sum of the first two. Therefore, S has the basis $\{(1, 2, 1, -1), (2, 0, 1, 1)\}$. Also, $\{(4, 4, 3, -1), (1, 0, 0, 0)\}$ is a basis of T. The union of these two sets gives the four-element set $\alpha_1 = (1, 2, 1, -1)$, $\alpha_2 = (2, 0, 1, 1)$, $\alpha_3 = (4, 4, 3, -1)$, $\alpha_4 = (1, 0, 0, 0)$, which spans $S + T$. We look for a maximal linearly independent subset of $\{\alpha_1, \alpha_2, \alpha_3, \alpha_4\}$. The set $\{\alpha_1\}$ is linearly independent. So is the set $\{\alpha_1, \alpha_2\}$. However, $\alpha_3 = 2\alpha_1 + \alpha_2$, so $\{\alpha_1, \alpha_2, \alpha_3\}$ is linearly dependent and, therefore, $S + T$ is spanned by $\{\alpha_1, \alpha_2, \alpha_4\}$. This last set is linearly independent, because α_4 is not a linear combination of $\{\alpha_1, \alpha_2\}$.

We display $\{\alpha_1, \alpha_2, \alpha_4\}$ as a basis of $S + T$ and note that $\dim (S + T) = 3$. Recall that $\dim S = \dim T = 2$, and compute that $\dim (S \cap T) = 1$. This means that any nonzero element of $S \cap T$ is a basis of $S \cap T$. We observed in the preceding paragraph that $\alpha_3 = 2\alpha_1 + \alpha_2 = (4, 4, 3, -1)$. Therefore, $\{(4, 4, 3, -1)\}$ is a basis of $S \cap T$.

We conclude the lesson with an application to systems of homogeneous linear equations. Consider the system

$$\begin{aligned} 2x + y + 5z + w &= 0 \\ x - 3y + z + w &= 0 \\ 2x + z - w &= 0 \end{aligned}$$

The system can be written in "vector" form:

$$x\begin{bmatrix} 2 \\ 1 \\ 2 \end{bmatrix} + y\begin{bmatrix} 1 \\ -3 \\ 0 \end{bmatrix} + z\begin{bmatrix} 5 \\ 1 \\ 1 \end{bmatrix} + w\begin{bmatrix} 1 \\ 1 \\ -1 \end{bmatrix} = \begin{bmatrix} 0 \\ 0 \\ 0 \end{bmatrix}$$

where the vectors $(2, 1, 2)$, $(1, -3, 0)$, etc., of R^3 are arranged vertically for aesthetic reasons. Since any four vectors of R^3 are linearly dependent, we know that the system has a nontrivial solution. The result is general.

A system of fewer than n homogeneous linear equations in n unknowns has a nontrivial solution.

4-5 Theorem

It is left for you to quote the appropriate theorem.

Summary of Topics in Lesson 4

The number of elements in a basis.

Lessons in Linear Algebra

Extending a linearly independent set to a basis in a finite-dimensional space.

A comparison of the spaces $S, T, S \cap T, S + T$, with special attention to their dimensions. Using the relation $\dim S + \dim T = \dim (S \cap T) + \dim (S + T)$ to examine the spaces.

Writing a system of linear equations in vector form.

Applying the results of the lesson to show that a system of fewer than n homogeneous equations in n unknowns must have a nontrivial solution.

Exercises

1. Extend the set $A = \{(1, 1, 0), (1, -1, 0)\}$ to a basis of R^3.

2. Extend the set $\alpha_1 = (1, 2, 3, 4)$, $\alpha_2 = (-1, 0, 1, 0)$ to a basis of R^4 by taking B as the standard basis and following the procedure suggested in the proof of Theorem 4-2.

3. Show that the real solutions of the equation $2x + 3y = 0$ form a subspace S of R^2, and find a basis of S.

4. Show that the real solutions (x, y, z) of the system $x + y - z = 0$, $2x + y + 3z = 0$ form a subspace S of R^3, and find a basis of S.

5. Let S and T be finite-dimensional subspaces of a linear space.
   (a) If $\dim S = \dim T$, does it follow that $S = T$?
   (b) If $S \subset T$, explain why $\dim S \leq \dim T$.
   (c) If $S \subset T$ and $\dim S = \dim T$, explain why $S = T$.

6. Let S be spanned by the set $\alpha_1 = (1, 1, 0, 0, 0)$, $\alpha_2 = (0, 0, 1, 1, 1)$, $\alpha_3 = (1, 1, 1, 1, 1)$ and let T be spanned by $\beta_1 = (2, 2, 3, 3, 3)$, $\beta_2 = (1, 0, 1, 0, 1)$, $\beta_3 = (0, 0, 0, 0, 1)$. Find a basis and the dimension for each of the spaces $S, T, S + T, S \cap T$.

7. Let S be the function space spanned by $\{\sin^2 x, \cos^2 x, \cos 2x\}$ and let T be spanned by $\{1, x, x^2\}$. Find a basis and the dimension for each of the spaces $S, T, S + T, S \cap T$.

8. Prove that if S is an r-dimensional subspace of the n-dimensional vector space V, then there is an $n-r$ dimensional subspace T such that $S \cap T = 0$ and $S + T = V$. (Hint: Let $\{\alpha_1, \ldots, \alpha_r\}$ be a basis of S. Extend to the basis $\{\alpha_1, \ldots, \alpha_r, \beta_1, \ldots, \beta_{n-r}\}$ of V.)
   Additional remark: In case $S \cap T = 0$ and $S + T = V$, we say V is the *direct* sum of S and T.

More on Bases, Spanning Sets, and Dimension

9. Show that if $A = \{\alpha_1, \alpha_2, \alpha_3, \alpha_4\}$ is a basis of V, then $B = \{\alpha_1 + \alpha_2 + \alpha_3 + \alpha_4, \alpha_2, \alpha_3, \alpha_4\}$ is a basis of V.

10. Let S be the solution space of the homogeneous equation $x + y + z + w = 0$, and let T be the solution space of the equation $x - y + z - w = 0$.
    (a) Explain why $S \cap T$ is the solution space of the system consisting of both equations.
    (b) Find a basis for each of the spaces S, T, $S + T$, $S \cap T$.

11. (a) Prove: If S and T are *subsets* of the linear space V, then $L(S \cap T) \subset L(S) \cap L(T)$.
    (b) Disprove by giving a counterexample: $L(S \cap T) = L(S) \cap L(T)$ for all subsets S, T.

12. (a) Let α_1 and α_2 be any two vectors in R^n, and let $\beta_1 = a_{11}\alpha_1 + a_{12}\alpha_2$, $\beta_2 = a_{21}\alpha_1 + a_{22}\alpha_2$, $\beta_3 = a_{31}\alpha_1 + a_{32}\alpha_2$. Prove that $\{\beta_1, \beta_2, \beta_3\}$ is linearly dependent. You may use Theorem 4-5.
    (b) Write out a generalization of the result proved in (a).

13. Let S_1, S_2, T_1, T_2 be four subspaces of a finite-dimensional vector space, and suppose dim $(S_1 \cap T_1) =$ dim $(S_2 \cap T_2)$, dim $(S_1 + T_1) =$ dim $(S_2 + T_2)$, dim $T_1 =$ dim T_2, and $S_1 \subset S_2$. Prove $S_1 = S_2$.

Remarks on Exercises

1. You may choose any third vector α_3 so that $A \cup \{\alpha_3\}$ is linearly independent. We can pick a vector $\alpha_3 = (0, 0, 1)$ from the standard basis to form with A the basis $\{(1, 1, 0), (1, -1, 0), (0, 0, 1)\}$.

2. $\{(1, 2, 3, 4), (-1, 0, 1, 0), (1, 0, 0, 0), (0, 1, 0, 0)\}$.

3. The solution set $S = \{(3t, -2t) \mid t \in R\}$ is the subspace spanned by the one element $(3, -2)$.

4. The given system is equivalent to the system $x + 4z = 0$, $-y + 5z = 0$. The solution space $\{(-4z, 5z, z) \mid z \in R\}$ has a basis of one element, $(-4, 5, 1)$.

5. (a) No.
   (b) A basis of S could be extended to a basis of T.
   (c) The extension of a basis of S to a basis of T is trivial, and any basis of S is a basis of T.

6. S has basis $\{\alpha_1, \alpha_2\}$, T has basis $\{\beta_1, \beta_2, \beta_3\}$, $S + T$ has basis $\{\alpha_1, \alpha_2, \beta_2, \beta_3\}$. By Theorem 4-4, dim $(S \cap T) = 1$. A basis of $S \cap T$ is $\{\beta_1\}$.

Lessons in Linear Algebra

7. A basis of S is $\{\sin^2 x, \cos^2 x\}$. A basis of T is $\{1, x, x^2\}$. A basis of $S + T$ is $\{\sin^2 x, \cos^2 x, x, x^2\}$. A basis of $S \cap T$ is $\{1\}$.

9. Either of two approaches suffices. Show B is linearly independent. Or, show each vector of A is a linear combination of B.

10. (b) The solution space of S is $\{(x, y, z, -x - y - z) \mid x, y, z \in R\}$ or $\{x(1, 0, 0, -1) + y(0, 1, 0, -1) + z(0, 0, 1, -1) \mid x, y, z \in R\}$. A basis of S is $\{(1, 0, 0, -1), (0, 1, 0, -1), (0, 0, 1, -1)\}$. Similarly, a basis of T is $\{(1, 0, 0, 1), (0, 1, 0, -1), (0, 0, 1, 1)\}$. A basis of $S + T$ is $\{(1, 0, 0, -1), (0, 1, 0, -1), (0, 0, 1, -1), (1, 0, 0, 1)\}$. A basis of $S \cap T$ is $\{(0, 1, 0, -1), (1, 0, -1, 0)\}$.

13. Use Theorem 4-4.

5

APPLICATIONS TO SYSTEMS OF LINEAR EQUATION

The study of systems of linear equations is important in all sciences, especially engineering, physics, and chemistry. In recent years the use of linear equations in social sciences such as psychology, economics, and sociology has made it desirable for social science students to include linear spaces in their plan of study.

In this lesson we will learn how to solve systems of linear equations. Remember that a system may have exactly one solution, more than one solution, or no solution. To see this consider three very simple systems:

(1) $x + y = 3, x - y = 1$.
(2) $x + y = 2$.
(3) $x + y = 2, 2x + 2y = 1$.

System (1) has exactly one solution (2, 1). System (2) has the solution set $\{(x, 2, -x)\}$ or $\{(0, 2) + x(1, -1)\}$, which contains many elements. System (3) has no solution, and is, therefore, called *inconsistent*. Of course systems (1) and (2) are both *consistent*.

To solve a problem all of its solutions must be found. *We either display the complete set of solutions or else show that there are no solutions.* Let us consider a general system of m linear equations in n unknowns,

$$\sum_{j=1}^{n} a_{ij}x_j = b_i, \quad i = 1, 2, \ldots, m$$

The system may be written

$$a_{11}x_1 + a_{12}x_2 + \cdots + a_{1n}x_n = b_1$$
$$a_{21}x_1 + a_{22}x_2 + \cdots + a_{2n}x_n = b_2$$
$$\vdots$$
$$a_{m1}x_1 + a_{m2}x_2 + \cdots + a_{mn}x_n = b_m$$

Lessons in Linear Algebra

Indeed if we keep the unknowns $x_1, \ldots, x_n$ in mind, we can easily visualize the system as the "matrix" (rectangular array)

$$M = \begin{bmatrix} a_{11} & a_{12} & \cdots & a_{1n} & b_1 \\ a_{21} & a_{22} & \cdots & a_{2n} & b_2 \\ & & \vdots & & \\ a_{m1} & a_{m2} & \cdots & a_{mn} & b_m \end{bmatrix}$$

There are many procedures for solving the system. We develop one which always works and is especially adaptable to computers. First we explain a new operation.

5-1 Definition

An *elementary operation* on a set $A = \{\alpha_1, \alpha_2, \ldots\}$ of vectors is either one of the following:

(1) multiplying a vector by a nonzero scalar
(2) adding to one vector a multiple of another
(3) swapping two vectors

Right now, we will use elementary operations in solving linear systems. Later we will perform elementary operations on other sets.

The rows of M, each an $(n + 1)$-tuple, represent the equations of the system. We plan to perform elementary operations on the rows of M.

5-2 Definition

Two systems of equations are *equivalent* if they have the same solution sets.

5-3 Theorem

Performing an elementary operation on a system of linear equations does not change the solution set.

Proof: This powerful fact is easy to prove. Think of the system as $E_1 = 0, E_2 = 0, \ldots, E_m = 0$. See Definition 5-1 and note that $E_i = 0$ if and only if $kE_i = 0$ (given $k \neq 0$). Similarly, system $E_1 + kE_2 = 0$, $E_2 = 0, \ldots, E_m = 0$ is satisfied if and only if $E_1 = 0, E_2 = 0, \ldots, E_m = 0$. The third case is trivial. ∎

Next we solve a system by applying elementary operations to the given system in order to change it into a form from which the solution

Applications to Systems of Linear Equation

set may be clearly displayed. The procedure used is called the *Gauss-Jordan method*.*

Example 1. Consider the system

$$\begin{aligned} x + y - z + w + u &= 5 \\ 2x + 3y + z - u &= -9 \\ 4x + 5y - z + 2w + u &= 1 \end{aligned}$$

or consider the following matrix, which is essentially the same as the system:

$$M = \begin{bmatrix} 1 & 1 & -1 & 1 & 1 & 5 \\ 2 & 3 & 1 & 0 & -1 & -9 \\ 4 & 5 & -1 & 2 & 1 & 1 \end{bmatrix}$$

Let R_1, R_2, R_3 represent the three rows (equations). We add to R_2 the vector $-2R_1$. This gives

$$\begin{bmatrix} 1 & 1 & -1 & 1 & 1 & 5 \\ 0 & 1 & 3 & -2 & -3 & -19 \\ 4 & 5 & -1 & 2 & 1 & 1 \end{bmatrix}$$

Then adding the appropriate multiple of R_1 to R_3 gives

$$\begin{bmatrix} 1 & 1 & -1 & 1 & 1 & 5 \\ 0 & 1 & 3 & -2 & -3 & -19 \\ 0 & 1 & 3 & -2 & -3 & -19 \end{bmatrix}$$

Next adding -1 times the second row to the third gives

$$\begin{bmatrix} 1 & 1 & -1 & 1 & 1 & 5 \\ 0 & 1 & 3 & -2 & -3 & -19 \\ 0 & 0 & 0 & 0 & 0 & 0 \end{bmatrix}$$

Finally, adding -1 times the second row to the first gives

$$\begin{bmatrix} 1 & 0 & -4 & 3 & 4 & 24 \\ 0 & 1 & 3 & -2 & -3 & -19 \\ 0 & 0 & 0 & 0 & 0 & 0 \end{bmatrix}$$

---

*Carl Friedrich Gauss (1777–1855), a German, has been called the Prince of Mathematics. His works touch almost every known branch of mathematics and have opened new areas still under study. He is especially remembered for his proof of the fundamental theorem of algebra — that every polynomial equation has a complex root.

The French mathematician Camille Jordan (1838–1922) was noted for his work in analysis and algebra.

Lessons in Linear Algebra

We have the system now in "reduced echelon" form, and all we have to do is look at the system and think. It is equivalent to the original system. We examine, therefore, the system

$$x - 4z + 3w + 4u = 24$$
$$y + 3z - 2w - 3u = -19$$

and note that the general solution (x, y, z, w, u) is given by

$$x = 24 + 4z - 3w - 4u$$
$$y = -19 - 3z + 2w + 3u$$

where z, w, u are arbitrary.

Also the solution set can be displayed as

$$\{(24 + 4z - 3w - 4u, -19 - 3z + 2w + 3u, z, w, u) \mid z, w, u \in R\}$$

or as

$$\{(24, -19, 0, 0, 0) + z(4, -3, 1, 0, 0) + w(-3, 2, 0, 1, 0)$$
$$+ u(-4, 3, 0, 0, 1) \mid z, w, u \in R\}$$

The reader should check that this general solution does indeed satisfy each of the original equations, regardless of the values of z, w, u.

The last matrix obtained in the example is called the "Hermite normal" form or "reduced echelon" form of the original matrix. Changing a system by elementary row operations can lead to many different "echelon" forms, but a given matrix has only one reduced echelon form (Hermite normal form), as we shall see.

5-4 Definition

A *matrix of type* (m, n) is a rectangular array of elements with m rows and n columns,

$$A = \begin{bmatrix} a_{11} & a_{12} & \ldots & a_{1n} \\ a_{21} & a_{22} & \ldots & a_{2n} \\ & & \vdots & \\ a_{m1} & a_{m2} & \ldots & a_{mn} \end{bmatrix}$$

We denote such a matrix by $A = [a_{ij}]$, or by $A = [a_{ij}]_{m,n}$ whenever we wish to give the type (m, n) of the matrix. A *real* matrix is one whose elements are real, and a *complex* matrix is one whose elements are complex.

Two matrices $A = [a_{ij}]$ and $B = [b_{ij}]$ are *equal* if they are of the same type and $a_{ij} = b_{ij}$ for all i, j.

Unless otherwise stated, each matrix we consider will be complex. Most of our examples will involve real matrices.

For example

$$\begin{bmatrix} 1 & \sqrt{2} & 3 \\ 1 & -1 & 0 \end{bmatrix}$$

is a real matrix of type (2, 3). The matrix

$$\begin{bmatrix} 1 + i \\ 3 \end{bmatrix}$$

is a complex matrix of type (2, 1). Of course every real matrix is also complex.

5-5 Definition Let A be a matrix of type m, n. A is said to be in *Hermite normal form** (or *reduced echelon form*) if

(1) the nonzero rows of A are at the top of the matrix, and the zero rows are at the bottom.
(2) the first nonzero entry in each nonzero row is 1, and occurs to the right of the first nonzero entry in the preceding row.
(3) the entry 1 referred to in (2) is the only nonzero entry in its column.

One needs examples in order to understand Definition 5-5.

Example 2. The matrix

$$A = \begin{bmatrix} 1 & 1 & 2 & 3 & 4 \\ 0 & 1 & 3 & -1 & 0 \\ 0 & 0 & 0 & 1 & 5 \\ 0 & 0 & 0 & 0 & 0 \end{bmatrix}$$

satisfies conditions (1) and (2) of Definition 5-5 but not (3). By elementary row operations on A we get

---

*Charles Hermite (1822–1901) was an outstanding French mathematician. His many works include a proof that e is transcendental. Also he is noted for his work in such diverse areas as quadratic forms, number theory, and solutions of algebraic equations.

He did not perform well as a young man in mathematics examinations. This may indicate something about his examiners.

Lessons in Linear Algebra

$$B = \begin{bmatrix} 1 & 0 & -1 & 0 & -16 \\ 0 & 1 & 3 & 0 & 5 \\ 0 & 0 & 0 & 1 & 5 \\ 0 & 0 & 0 & 0 & 0 \end{bmatrix}$$

a matrix which satisfies all three conditions and is in Hermite normal form. The matrix A is in echelon form but not reduced echelon (Hermite normal) form.

Example 3. We reduce the matrix

$$A = \begin{bmatrix} 1 & -3 & 1 & 0 & 2 \\ 3 & -9 & 4 & 0 & 5 \\ 2 & -6 & 5 & 1 & 9 \\ 2 & -6 & 1 & 0 & 5 \end{bmatrix}$$

by performing three well chosen elementary row operations, specifically by adding to the last three rows appropriate multiples of the first row. This gives

$$B = \begin{bmatrix} 1 & -3 & 1 & 0 & 2 \\ 0 & 0 & 1 & 0 & -1 \\ 0 & 0 & 3 & 1 & 5 \\ 0 & 0 & -1 & 0 & 1 \end{bmatrix}$$

The first two columns of B are acceptable. The element 1 in the second row and third column must be the only nonzero element in column 3. This motivates us to add certain multiples of row 2 to the other rows in order to produce the desired zeros without disturbing the first two columns.

We get then the Hermite normal form

$$C = \begin{bmatrix} 1 & -3 & 0 & 0 & 3 \\ 0 & 0 & 1 & 0 & -1 \\ 0 & 0 & 0 & 1 & 8 \\ 0 & 0 & 0 & 0 & 0 \end{bmatrix}$$

For a given matrix A, there are many echelon forms, but there is just one Hermite normal form. In solving a linear system it is not always desirable to get the system in Hermite normal form. For example, the system $x + y = 4$, $x - y = 2$ may be solved quickly by simply adding the equations mentally. Finding the Hermite normal form requires more effort,

$$\begin{bmatrix} 1 & 1 & 4 \\ 1 & -1 & 2 \end{bmatrix} \sim \begin{bmatrix} 1 & 1 & 4 \\ 0 & -2 & -2 \end{bmatrix} \sim \begin{bmatrix} 1 & 1 & 4 \\ 0 & 1 & 1 \end{bmatrix} \sim \begin{bmatrix} 1 & 0 & 3 \\ 0 & 1 & 1 \end{bmatrix}$$

but we do emerge with the solution $x = 3$, $y = 1$.

Applications to Systems of Linear Equation

The advantage in reducing a system to Hermite normal form becomes apparent to one who seeks a definite systematic way of solving any system of m linear equations in n unknowns, a method which gets to the bottom of things whether or not the system is consistent. The method displays in a *unique* way all of the solutions.

As one reduces a given matrix to Hermite normal form, he has options as to just what elementary row operations he will use. The fact that the end product is unique may be surprising.

5-6 Definition

Two matrices are *row-equivalent* if one can be changed into the other by a finite number of elementary row operations.

It is easy to verify that *row-equivalence* of matrices is what one expects in an "equivalence" relation. It is a reflexive relation in that any matrix is row-equivalent to itself. It is a symmetric relation in that if A is row-equivalent to B, then B is row-equivalent to A. Finally, the relation is transitive in that if A is row-equivalent to B and B is row-equivalent to C, then A is row-equivalent to C.

5-7 Theorem

Let A be any matrix. There is a unique matrix B which is row-equivalent to A and is also in Hermite normal form.

Proof: We proceed by mathematical induction. Let A be of type (m, n). If A has just one column ($n=1$), then the theorem holds.

Next we take as our induction hypothesis that the theorem holds for each matrix with $n-1$ columns and attempt to conclude that the theorem holds when A has n columns. Accordingly we assume that elementary row operations have been performed on A until its first $n-1$ columns are in unique Hermite normal form. Then further elementary row operations bring the entire matrix into Hermite normal form without affecting the first $n-1$ columns.

It is left for you to argue that if the last column so obtained were not unique, then we could get two Hermite normal forms representing inequivalent systems of linear equations. This violates Theorem 5-3. ∎

Example 4. Consider the electrical circuit (see Figure 5-1) where two 12-volt cells are connected with three resistors of 4 ohms, 7 ohms, and

Lessons in Linear Algebra

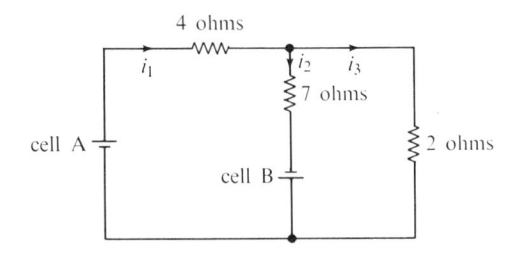

FIGURE 5-1

2 ohms. The currents at various points are labeled i_1, i_2, i_3 and by Ohm's law* satisfy

$$\begin{aligned} i_1 - i_2 - i_3 &= 0 \\ 4i_1 + 7i_2 &= 24 \\ 4i_1 + 2i_3 &= 12 \\ -7i_2 + 2i_3 &= -12 \end{aligned}$$

The Hermite normal form is (you should check this)

$$\begin{bmatrix} 1 & 0 & 0 & \frac{66}{25} \\ 0 & 1 & 0 & \frac{48}{25} \\ 0 & 0 & 1 & \frac{18}{25} \\ 0 & 0 & 0 & 0 \end{bmatrix}$$

and shows that $i_1 = 66/25$, $i_2 = 48/25$, $i_3 = 18/25$.

Example 5. A plant manufacturing three kinds of tools reported that its average daily production was 150 tools. This daily production figure included screwdrivers which cost $1 each and retailed at $2, wrenches which cost $2 each and retailed at $3, and shears which cost $3 each and retailed at $4. For an average day the cost was reported as $320, the total retail value as $500. Find how many of each were produced.

---

*The systematic and precise statement of laws governing currents and voltages related to junctions and loops in electric circuits is usually credited to the German physicist Gustav Robert Kirchhoff (1824–1887). In fact they are called "Kirchhoff's Laws."

Applications to Systems of Linear Equation

Solution: Let x_1 = number of screwdrivers produced, etc. We have the system

$$x_1 + x_2 + x_3 = 150$$
$$x_1 + 2x_2 + 3x_3 = 320$$
$$2x_1 + 3x_2 + 4x_3 = 500$$

or, in Hermite normal form (check it)

$$\begin{bmatrix} 1 & 0 & -1 & 0 \\ 0 & 1 & 2 & 0 \\ 0 & 0 & 0 & 1 \end{bmatrix}$$

This system

$$x_1 \quad\quad - x_2 = 0$$
$$x_1 + 2x_2 = 0$$
$$0x_1 + 0x_2 + 0x_3 = 1$$

is inconsistent, because of the third equation. Therefore, the report is incorrect. In fact, the inconsistency can be noted before the reduction to Hermite normal form is complete.

Summary of Topics in Lesson 5

Systems of linear equations. Consistent systems, inconsistent systems.

The matrix of a system of linear equations.

Elementary operations on a set of vectors.

Elementary row operations on a matrix. Elementary operations on a system of linear equations.

Equivalent systems of linear equations.

Changing a system of linear equations to Hermite normal form (reduced echelon form) and displaying the solution set.

Uniqueness of the Hermite normal form.

Exercises

1. Which of the following are in Hermite normal form?

   (a) $\begin{bmatrix} 0 & 0 & 2 & 0 & 3 \\ 0 & 1 & -4 & 0 & 4 \\ 0 & 0 & 0 & 1 & 0 \\ 0 & 0 & 0 & 0 & 0 \end{bmatrix}$
   (b) $\begin{bmatrix} 1 & 0 & 1 & 0 & 0 \\ 0 & 1 & 0 & 0 & 0 \\ 0 & 0 & 1 & 0 & 0 \\ 0 & 0 & 0 & 1 & 0 \end{bmatrix}$

57

(c) $\begin{bmatrix} 1 & 0 & 0 & 0 \\ 0 & 1 & 0 & 0 \\ 0 & 0 & 1 & 0 \\ 0 & 0 & 0 & 1 \end{bmatrix}$ (d) $\begin{bmatrix} 1 & 2 & 0 & 3 & 0 \\ 0 & 0 & 1 & 4 & 0 \\ 0 & 0 & 0 & 0 & 1 \\ 0 & 0 & 0 & 0 & 0 \end{bmatrix}$

2. Solve each of the following systems by reducing the matrix to Hermite normal form:

(a) $7x_1 + x_2 + 3x_3 = 16$
$2x_1 + x_2 + x_3 = 4$
$x_1 - x_2 - x_3 = 2$

(b) $3x_1 - x_2 + x_3 = 8$
$7x_1 + x_2 + 3x_3 = 16$
$2x_1 + x_2 + x_3 = 4$

(c) $3x_1 - x_2 + x_3 = 8$
$7x_1 + x_2 + 3x_3 = 16$
$2x_1 + x_2 + x_3 = 5$

(d) $x_1 - x_2 + x_3 - x_4 = 4$
$3x_1 - 4x_2 + 2x_3 - 2x_4 = 5$
$5x_1 - 10x_2 + 2x_3 - 2x_4 = -1$

(e) $x_1 + x_2 + x_3 + x_4 + x_5 = 3$
$2x_1 + 3x_2 - x_3 + x_5 = 7$
$3x_1 + 4x_2 + x_4 + 2x_5 = 10$
$4x_1 + 5x_2 + x_3 + 2x_4 + 3x_5 = 13$

(f) $x_1 + x_2 + x_3 + x_4 + x_5 = 0$
$2x_1 + 3x_2 - x_3 + x_5 = 0$

3. Definition 5-1 gives three types of elementary operations. Show that type (3) can be achieved by repeated applications of types (1) and (2).

4. Given that $a_{ij} = i + j$ for all $i = 1, 2, j = 1, 2, 3$, write out and solve the system

$$\sum_{j=1}^{3} a_{ij} x_j = i^2, \quad i = 1, 2$$

5. Let

$$A = \begin{bmatrix} 1 & 2 & 3 & 1 \\ -1 & -1 & 0 & 2 \\ 1 & 3 & 6 & 4 \end{bmatrix}$$

(a) Find a basis of the subspace of R^4 spanned by the rows of A.

(b) Find a basis of the subspace of R^3 spanned by the columns of A, $(1, -1, 1)$, $(2, -1, 3)$, $(3, 0, 6)$, $(1, 2, 4)$.
(c) Find B, the Hermite normal form of A.
(d) Find a basis for the subspace spanned by the rows of B.
(e) Find a basis for the subspace spanned by the columns of B.

6. If cell B of Example 4 is reversed, then the equations become

$$\begin{aligned} i_1 - i_2 - i_3 &= 0 \\ 4i_1 + 7i_2 &= 0 \\ 4i_1 + 2i_3 &= 12 \\ -7i_2 + 2i_3 &= 12 \end{aligned}$$

Solve the system.

7. Let $A = \{\alpha_1, \alpha_2, \ldots\}$ be a set of vectors, and let $B = \{\beta_1, \beta_2, \ldots\}$ be a set obtained by applying a finite number of elementary operations to A. Prove that A is linearly independent if and only if B is linearly independent. (Suggestion: Consider just one elementary operation.)

Remarks on Exercises

1. (c) and (d) are in Hermite normal form.
2. (a) $(2, -1, 1)$.
   (b) $\{(12/5, -4/5, 0) + x_3(-2/5, -1/5, 1) \mid x_3 \in R\}$.
   (c) Inconsistent.
   (d) $\{(-3, 0, 7, 0) + x_4(0, 0, 1, 1) \mid x_4 \in R\}$.
   (e) $\{(2, 1, 0, 0, 0) + x_3(-4, 3, 1, 0, 0) + x_4(-3, 2, 0, 1, 0) + x_5(-2, 1, 0, 0, 1) \mid x_3, x_4, x_5 \in R\}$.
   (f) The subspace spanned by $\{(-4, 3, 1, 0, 0), (-3, 2, 0, 1, 0), (-2, 1, 0, 0, 1)\}$.
4. $\{(8, -5, 0) + x_3(1, -2, 1) \mid x_3 \in R\}$.
5. (a) Any two rows of A will do.
   (b) Any two columns of A.
   (c) $B = \begin{bmatrix} 1 & 0 & -3 & -5 \\ 0 & 1 & 3 & 3 \\ 0 & 0 & 0 & 0 \end{bmatrix}$.
   (d) The first two rows of B.
   (e) The first two columns of B.
6. $(i_1, i_2, i_3) = (1/25)(42, -24, 66)$.

6

INNER PRODUCTS

In this lesson we extend the concepts of inner product, distance, and angle to linear spaces of dimension greater than 3. The fact that the study of linear spaces has roots in geometry is obvious from the use of words such as "length," "distance," "angle," and "projection" in spaces where things cannot be visualized as in three-space (or two-space).

When a new definition is made, the reader should compare it with the geometric meaning in two- or three-space. This can help him to remember the definition and to anticipate results. There is a possible bad effect in referring always to three-space. One may be misled into using visual arguments, perhaps valid in three-space, but not valid in n-space. The idea is to consider the definitions as they are written and to make conclusions by taking algebraic or analytic steps clearly valid in the space. As you develop confidence in your skills in n-space you will notice that proofs are often just as easy in n-space as in three-space. Also, you will come to see that it is relevant, for example, to talk of the function from the space spanned by $\{1, \sin x, \cos x, \sin 2x, \cos 2x\}$ which is nearest to a given function $f(x)$.

If you need a quick review of the dot product in two- or three-space, see the preliminary lesson.

6-1 Definition

Let V be a real vector space. V is said to be a *Euclidean space* (or *real inner product* space)* if for each pair of vectors α, β in V there is a scalar (α, β) called an *inner product*, such that for all vectors α, β, γ and scalars a the following four conditions hold:

---

*We will use interchangeably the expressions "Euclidean space" and "real inner product space," whether the dimension of the space is finite or infinite. Similarly, a complex inner product space is sometimes called a "unitary space."

Inner Products

(1) $(\alpha, \alpha) \geq 0$; $(\alpha, \alpha) = 0$ if and only if $\alpha = 0$.
(2) $(\alpha, \beta) = (\beta, \alpha)$.
(3) $(\alpha, \beta + \gamma) = (\alpha, \beta) + (\alpha, \gamma)$.
(4) $(a\alpha, \beta) = a(\alpha, \beta)$.

There is a slight redundancy in the above, since (4) implies part of (1) when $a = 0$.

The dot product in two- or three-space is an example of an inner product. Indeed the dot product can be extended into R^n.

6-2 Definition

For $\alpha = (a_1, \ldots, a_n)$ and $\beta = (b_1, \ldots, b_n)$, vectors in R^n, we define the *dot product by*

$$\alpha \cdot \beta = \sum_{i=1}^{n} a_i b_i$$

The reader should check that the dot product so defined is an inner product, that is, that the dot product satisfies all four conditions of Definition 6-1. As a start we consider condition (1). Note that

$$\alpha \cdot \alpha = \sum_{i=1}^{n} a_i^2 \geq 0 \quad \text{and} \quad \alpha \cdot \alpha = 0 \quad \text{if and only if } \alpha = 0$$

The dot product is but one example of an inner product. We now consider others.

Example 1. Let $V = C(a, b)$, the linear space of all real-valued functions which are continuous on the closed interval $[a, b]$, and take the inner product of two functions as

$$(f, g) = \int_a^b f(x)g(x)\, dx$$

Recall that from elementary calculus we can then express (f, g) as $\lim \sum f(x_i)g(x_i)\Delta x_i$ and are reminded by this of the dot Product. Using the above on the interval $[0, 1]$ gives, for example,

$$(x, x^2) = \int_0^1 x^3\, dx = \frac{1}{4}$$

$$(\sin x, \cos x) = \int_0^1 \sin x \cos x\, dx = \frac{1}{2}\sin^2 1$$

$$(1, 2x - 1) = \int_0^1 (2x - 1)\, dx = 0$$

Changing the interval changes the inner product.

To show that the space $C(a, b)$ of functions continuous on $[a, b]$ is a real inner product space we must check that the inner product defined by $\int_a^b f(x)g(x)\, dx$ satisfies all four conditions required of an inner product by Definition 6-1. Begin with condition (1).

$$(f, f) = \int_a^b [f(x)]^2\, dx \geq 0$$

because the integrand $[f(x)]^2$ is nonnegative on $[a, b]$. The other details are left to the student. You only need to quote the appropriate facts from elementary calculus.

The next project is to define length for vectors in a real inner product space. A classical inequality, due to Cauchy and Schwarz, is fundamental to our development of length.

6-3 Theorem (Schwarz inequality.) The inner product in a Euclidean space satisfies

$$(\alpha, \beta)^2 \leq (\alpha, \alpha)(\beta, \beta)$$

for all vectors α, β.

Proof: The proof is immediate if either α or β is 0. Let α and β be nonzero and consider $a\alpha + \beta$, where a is an arbitrary scalar. Apply Definition 6-1 to get

$$0 \leq (a\alpha + \beta, a\alpha + \beta) \quad \text{[part (1) of Definition 6-1]}$$
$$0 \leq a^2(\alpha, \alpha) + 2a(\alpha, \beta) + (\beta, \beta) \quad \text{[parts (2), (3), (4)]}$$

The preceding right member is a quadratic expression in a, with leading coefficient (α, α) positive. Since the expression is nonnegative for all a, the discriminant is negative or zero. That is,

$$4(\alpha, \beta)^2 - 4(\alpha, \alpha)(\beta, \beta) \leq 0$$

Hence we have the Schwarz inequality

$$(\alpha, \beta)^2 \leq (\alpha, \alpha)(\beta, \beta) \quad \blacksquare$$

6-4 Corollary

$(\alpha, \beta)^2 = (\alpha, \alpha)(\beta, \beta)$ if and only if $\{\alpha, \beta\}$ is linearly dependent.

Proof: See the proof of the theorem. We have $(\alpha, \beta)^2 = (\alpha, \alpha)(\beta, \beta)$ if and only if there is a scalar a such that $(a\alpha + \beta, a\alpha + \beta) = 0$ or if and only if $a\alpha + \beta = 0$. ∎

Let us illustrate the Schwarz inequality. In R^3 let $\alpha = (-1, 1, 1)$, $\beta = (1, -2, -3)$. Then $(\alpha \cdot \beta)^2 = (-6)^2 = 36$, $(\alpha \cdot \alpha) = 3$, $(\beta \cdot \beta) = 14$, and the Schwarz inequality predicts that $36 \leq 42$.

In $C(a, b)$, the space of functions continuous on $[a, b]$, the Schwarz inequality reveals that for all functions f and g,

$$\left[\int_a^b f(x)g(x)\,dx\right]^2 \leq \int_a^b [f(x)]^2\,dx \int_a^b [g(x)]^2\,dx$$

6-5 Definition

Let V be a Euclidean space with inner product (α, β). The *length* of α (or *norm* of α) is denoted $\|\alpha\|$ and is defined by

$$\|\alpha\| = \sqrt{(\alpha, \alpha)}$$

For example, in R^4 the dot product gives for $\alpha = (1, 1, 1, 2)$ the length $\|\alpha\| = 7$. In $C(0, 1)$, the usual inner product gives for $\alpha = \sqrt{2x}$,

$$\|\alpha\| = \sqrt{\int_0^1 2x\,dx} = 1$$

Such a vector α is given a special name.

6-6 Definition

A vector of norm 1 is called a *unit* vector.

In any real inner product space, the norm satisfies elementary properties, regardless of which inner product is used, due to the properties met by any inner product.

6-7 Theorem

Let V be a Euclidean space with inner product (α, β). Then for all α, β, a:

(1) $\|\alpha\| \geq 0$, $\|\alpha\| = 0$ if and only if $\alpha = 0$.
(2) $\|a\alpha\| = |a|\,\|\alpha\|$.
(3) $|(\alpha, \beta)| \leq \|\alpha\|\,\|\beta\|$. (The Schwarz inequality again.)
(4) $\|\alpha + \beta\| \leq \|\alpha\| + \|\beta\|$. (The "triangle" inequality.)

Lessons in Linear Algebra

Proof: Parts (1) and (2) follow from Definition 6-1 of inner product. Recall that $(\alpha, \alpha)^2 \geq 0$ and $(\alpha, \alpha) = 0$ if and only if $\alpha = 0$. This gives $\|\alpha\|^2 \geq 0$ and $\|\alpha\|^2 = 0$ if and only if $\alpha = 0$. Also $\|a\alpha\| = \sqrt{(a\alpha, a\alpha)} = \sqrt{a^2(\alpha, \alpha)} = |a| \|\alpha\|$ from properties (2) and (3) of Definition 6-1. Part (3) is merely a restatement of the Schwarz inequality. The proof of (4) is left for you. ∎

We conclude this lesson by introducing the concept of distance in a real inner product space.

6-8 Definition

Let α and β be two vectors in a real inner product space V. The *distance from α to β*, denoted $d(\alpha, \beta)$ is defined by

$$d(\alpha, \beta) = \|\alpha - \beta\|$$

To illustrate, in R^4 let $\alpha = (1, 1, 1, 2)$, $\beta = (0, 1, 2, 3)$. Then $d(\alpha, \beta) = \|\alpha - \beta\| = \|(1, 0, -1, -1)\| = \sqrt{3}$. In $C(0, 1)$, let $\alpha = x$, $\beta = x^2$. Then

$$d(\alpha, \beta) = \|x - x^2\| = \left[\int_0^1 (x - x^2)^2\, dx\right]^{1/2} = \sqrt{\frac{1}{3} - \frac{1}{2} + \frac{1}{5}} = \sqrt{\frac{1}{30}}$$

The distance defined above behaves nicely and gives a "metric" space, because it meets the conditions specified in the following theorem.

6-9 Theorem

In a real inner product space, the distance given by $d(\alpha, \beta) = \|\alpha - \beta\|$ satisfies

(1) $d(\alpha, \beta) \geq 0$, $d(\alpha, \beta) = 0$ if and only if $\alpha = \beta$.
(2) $d(\alpha, \beta) = d(\beta, \alpha)$ for all α, β.
(3) $d(\alpha, \beta) + d(\beta, \gamma) \geq d(\alpha, \gamma)$ for all α, β, γ. (The "triangle" inequality again.)

Proof: We have (1) because $\|\alpha - \beta\| \geq 0$ and $\|\alpha - \beta\| = 0$ if and only if $\alpha - \beta = 0$, from Theorem 6-7, part (1); and $d(\alpha, \beta) = d(\beta, \alpha)$ follows because $\|\alpha - \beta\| = \|\beta - \alpha\|$ from Theorem 6-7, part (2). To prove $d(\alpha, \beta) + d(\beta, \gamma) \geq d(\alpha, \gamma)$ use the triangle inequality (4) of Theorem 6-7 in the form $\|(\alpha - \beta) + (\beta - \gamma)\| \leq \|\alpha - \beta\| + \|\beta - \gamma\|$ and get $\|\alpha - \gamma\| \leq \|\alpha - \beta\| + \|\beta - \gamma\|$. ∎

Optional: Inner Products in Complex Spaces

The results of the lesson can be extended to complex vector spaces. A summary follows, with most of the details omitted. However, hints are provided for the more difficult points.

6-10 Definition

A complex vector space V is called a *complex inner product space* (or *unitary* space) if it has an inner product (α, β) satisfying the following conditions for all vectors α, β, γ and scalars a.

(1) $(\alpha, \alpha) \geq 0$, $(\alpha, \alpha) = 0$ if and only if $\alpha = 0$.
(2) $(\alpha, \beta) = \overline{(\beta, \alpha)}$, the complex conjugate of (α, β).
(3) $(\alpha, \beta + \gamma) = (\alpha, \beta) + (\alpha, \gamma)$.
(4) $(a\alpha, \beta) = a(\alpha, \beta)$.

For an example of a suitable inner product in C^2 let $\alpha = (a_1, a_2)$, $\beta = (b_1, b_2)$ and define $(\alpha, \beta) = a_1\bar{b}_1 + a_2\bar{b}_2$. (This can be extended in the obvious way to C^n.) If $\alpha = (1, i)$ and $\beta = (2 + i, i)$, then $(\alpha, \beta) = 1(2 - i) + i(-i) = 3 - i$. Also $(\beta, \beta) = (2 + i)(2 - i) + i(-i) = 6$.

6-11 Theorem

In a complex inner product space for all α, β, γ, b:

(1) $(\alpha, b\beta) = \bar{b}(\alpha, \beta)$.
(2) $(\beta + \gamma, \alpha) = (\beta, \alpha) + (\gamma, \alpha)$.
(3) $|(\alpha, \beta)|^2 \leq (\alpha, \alpha)(\beta, \beta)$. (Schwarz inequality)

Proof: In (1) and (2) remember that all complex numbers satisfy the conditions $\overline{ab} = \bar{a}\,\bar{b}$, $\overline{a+b} = \bar{a} + \bar{b}$, $|a| = \sqrt{a\,\bar{a}}$, etc. Then (1) follows from $(\alpha, b\beta) = \overline{(b\beta, \alpha)} = \overline{b(\beta, \alpha)} = \bar{b}\overline{(\beta, \alpha)} = \bar{b}(\alpha, \beta)$. Condition (2) is also easy to prove. Try it.

The Schwarz inequality is harder. Take α and β nonzero and consider

$$0 \leq (a\alpha + b\beta, a\alpha + b\beta)$$
$$0 \leq a\bar{a}(\alpha, \alpha) + a\bar{b}(\alpha, \beta) + b\bar{a}(\beta, \alpha) + b\bar{b}(\beta, \beta)$$

which holds for all a, b, α, β. Choose $b = (\alpha, \alpha) = \bar{b}$ and obtain

$$0 \leq a\bar{a}(\alpha, \alpha) + a(\alpha, \alpha)(\alpha, \beta) + \bar{a}(\alpha, \alpha)(\beta, \alpha) + (\alpha, \alpha)^2 (\beta, \beta)$$
$$0 \leq a\bar{a} + a(\alpha, \beta) + \bar{a}(\beta, \alpha) + (\alpha, \alpha)(\beta, \beta)$$

Finally, choose $a = -\overline{(\alpha, \beta)} = -(\beta, \alpha)$ and get

$$0 \leq |(\alpha, \beta)|^2 - |(\alpha, \beta)|^2 - |(\alpha, \beta)|^2 + (\alpha, \alpha)(\beta, \beta)$$

and conclude that $|(\alpha, \beta)|^2 \leq (\alpha, \alpha)(\beta, \beta)$. ∎

Lessons in Linear Algebra

We define the *norm* of α again by $\|\alpha\| = \sqrt{(\alpha, \alpha)}$. Then Theorem 6-7 holds. The results on distance, including Theorem 6-9 also hold for a complex Euclidean space.

Summary of Topics in Lesson 6

Inner products and real inner product spaces (Euclidean spaces).

Properties of the inner product.

The dot product.

An inner product in a function space.

The Schwarz inequality.

The norm (length) of a vector. Properties including the triangle inequality.

Distance between two vectors. Properties.

Optional section on complex inner product spaces (unitary spaces).

Exercises

1. In R^4, let $\alpha = (-1, 1, 1, 1)$, $\beta = (1, -1, -2, 1)$, $\gamma = (1, -1, -3, 1)$. Using the dot product as the inner product
   (a) illustrate the Schwarz inequality $|\alpha \cdot \beta| \leq \|\alpha\| \|\beta\|$ by computing both members.
   (b) verify the triangle inequality $\|\alpha + \beta\| \leq \|\alpha\| + \|\beta\|$.
   (c) verify the triangle inequality $d(\alpha, \beta) + d(\beta, \gamma) \geq d(\alpha, \gamma)$.
   (d) find three unit vectors which are multiples of α, β, γ, respectively.

2. In $C(-1, 1)$, the linear space of functions continuous on $[-1, 1]$, take as the inner product $(f, g) = \int_{-1}^{1} f(x)g(x) \, dx$.
   (a) Let $f(x) = 1$ and $g(x) = x^2$. Verify the Schwarz inequality $(f, g)^2 \leq (f, f)(g, g)$ and the triangle inequality $\|f + g\| \leq \|f\| + \|g\|$.
   (b) Compute $d(f, g)$, given that $f(x) = x^2$, and $g(x) = x^3$.

3. In a real inner product space, prove that for all α, β:
   (a) $(\alpha, \beta) = 0$ if and only if $\|\alpha + \beta\| = \|\alpha - \beta\|$. What theorem about parallelograms is indicated?
   (b) $\|\alpha + \beta\|^2 + \|\alpha - \beta\|^2 = 2[\|\alpha\|^2 + \|\beta\|^2]$. What theorem about parallelograms is indicated?

(c) $(\alpha, \beta) = 0$ if and only if $\|\alpha + \beta\|^2 = \|\alpha\|^2 + \|\beta\|^2$. Interpret in the plane.

(d) $(\alpha, \beta) = (1/2)[\|\alpha + \beta\|^2 - \|\alpha\|^2 - \|\beta\|^2]$. This shows that the inner product can be expressed in terms of norms.

4. In $C(0, 1)$ find the constant a such that ax is nearest to x^2 in the sense that $d(ax, x^2)$ is a minimum. Sketch a graph. (Use the inner product $(f, g) = \int_0^1 f(x)g(x)\, dx$.)

5. Let $p_1, p_2, \ldots, p_n$ be positive numbers, and let $\alpha = (a_1, a_2, \ldots, a_n)$, $\beta = (b_1, \ldots, b_n)$ be vectors of R^n. Show that $(\alpha, \beta) = \sum_{i=1}^{n} p_i a_i b_i$ is an inner product.

6. Let $p(x)$ be any function which is positive on the interval $[a, b]$ and define for the space $C(a, b)$, $(f, g) = \int_a^b p(x)f(x)g(x)\, dx$. Show that (f, g) is an inner product.

7. Let P be the space of all real polynomials, and define $(f, g) = \int_0^\infty e^{-x} f(x)g(x)\, dx$.

(a) Show that the integral converges for all polynomials and is, therefore, an acceptable inner product. Hint: Show $\int_0^\infty e^{-x} \cdot x^n\, dx = n!$ by induction.

(b) Compute (x^m, x^n).

(c) Find the norm of x.

(d) For $f(x) = (x + 2)^2$ and $g(x) = x + 1$, compute (f, g).

8. (a) Use the properties in Definition 6-1 to prove the right distributive law $(\beta + \gamma, \alpha) = (\beta, \alpha) + (\gamma, \alpha)$.

(b) Prove the triangle inequality (4) in Theorem 6-7. Start with $\|\alpha + \beta\|^2 = (\alpha + \beta, \alpha + \beta)$. You may use the Schwarz inequality at the appropriate time.

9. Let V be the set of all square convergent infinite sequences of real numbers.* That is, a typical vector of V is the infinite sequence $\{x_1, x_2, \ldots\}$ for which the infinite series $x_1^2 + x_2^2 + \ldots$ converges.

---

*This space is called a "Hilbert" space. The German mathematician David Hilbert (1862–1943) is noted for his work in foundations of geometry and also for his ability to formulate the great problems of mathematics.

Lessons in Linear Algebra

For the two sequences $\alpha = \{x_i\}$ and $\beta = \{y_i\}$, define $(\alpha, \beta) = \sum_{i=1}^{\infty} x_i \bar{y}_i$, an infinite series.

(a) Estimate $\sum_{i=1}^{n} |x_i y_i|$ by using the Schwarz inequality.
(b) Show that the series defining (α, β) converges absolutely and is an inner product.
(c) Compute (α, β) if $x_i = 1/i$ and $y_i = 1/(i-1)!$.
(d) Compute (α, β) if $x_i = 1/i$ and $y_i = 1/i + 1$.

Optional Exercises

10. In C^2, for typical vectors $\alpha = (a_1, a_2)$, $\beta = (b_1, b_2)$, define the inner product by $(\alpha, \beta) = a_1 \bar{b}_1 + a_2 \bar{b}_2$. For $\alpha = (1 + 2i, 3 - i)$ and $\beta = (4i, 3)$, compute (α, β), (β, α), $\|\alpha\|$, $\|\beta\|$, and verify the Schwarz inequality $|(\alpha, \beta)| \leq \|\alpha\| \|\beta\|$.

11. Prove part (2) of Theorem 6-11.

12. Show that in a complex inner product space. $(a\alpha + b\beta, c\gamma + d\delta) = a\bar{c}(\alpha, \gamma) + a\bar{d}(\alpha, \delta) + b\bar{c}(\beta, \gamma) + b\bar{d}(\beta, \delta)$.

13. Show that in a complex inner product space, for all α, β, $4(\alpha, \beta) = \|\beta + \alpha\|^2 - \|\beta - \alpha\|^2 - i\|\beta + i\alpha\|^2 + i\|\beta - i\alpha\|^2$. This is the complex equivalent of exercise 3d and shows that the inner product can be expressed in terms of the norm.

Remarks on Exercises

1. (a) $\alpha \cdot \beta = -3$, $\|\alpha\| = 2$, $\|\beta\| = \sqrt{7}$.
   (b) $\|\alpha + \beta\| = \sqrt{5} \leq 2 + \sqrt{7}$.
   (c) $d(\alpha, \beta) + d(\beta, \gamma) \geq d(\alpha, \gamma)$ because $\sqrt{17} + 1 \geq \sqrt{24}$.
   (d) $(1/2)\alpha$, $(1/\sqrt{7})\beta$, $(1/\sqrt{12})\gamma$.

2. (a) $(f, g) = (2/3)$, $(f, f) = 2$, $(g, g) = (2/5)$. $\|f + g\| \leq \|f\| + \|g\|$ because $\sqrt{56/15} \leq \sqrt{2} + \sqrt{2/5}$.

3. (a) Start with $(\alpha + \beta, \alpha + \beta) = (\alpha - \beta, \alpha - \beta)$. A parallelogram is a rectangle if and only if its diagonals are equal.
   (b) The sum of the squares of the four sides of a parallelogram equals the sum of the squares of the diagonals.
   (c) The Pythagorean theorem with converse.
   (d) Start with the right member and change each norm to an inner product.

4. The distance $d(ax, x^2) = [(a^2/3) - (a/2) + (1/5)]^{1/2}$, a function of a which attains its minimum when $a = 3/4$. The line $y = ax$ is chosen with $a = 3/4$ so that the integral of the square of the difference between $y = ax$ and $y = x^2$ on the interval $[0, 1]$ is a minimum. We have an infinite analogue of a least squares approximation.

7. (b) $(m + n)!$
   (c) $\sqrt{2}$.
   (d) $(f, g) = (x^2 + 4x + 4, x + 1) = (x^2, x) + (x^2, 1) + \ldots$.
       Now use (b) to get $(f, g) = 28$.

8. (b) $\|\alpha + \beta\|^2 = (\alpha + \beta, \alpha + \beta)$ (definition of norm)
   $= (\alpha, \alpha) + (\alpha, \beta) + (\beta, \alpha) + (\beta, \beta)$ (distributive laws)
   $= (\alpha, \alpha) + 2(\alpha, \beta) + (\beta, \beta)$
   $\leq \|\alpha\|^2 + 2\|\alpha\|\,\|\beta\| + \|\beta\|^2$ (Schwarz)
   $\leq (\|\alpha\| + \|\beta\|)^2$

9. (a) Applying the Schwarz inequality to the vectors $(|x_1|, |x_2|, \ldots, |x_n|), (|y_1|, |y_2|, \ldots, |y_n|)$ of R^n gives $\sum_{i=1}^{n} |x_i y_i| = \sum_{i=1}^{n} |x_i||y_i|$
   $\leq \sum_{i=1}^{n} x_i^2 \sum_{i=1}^{n} y_i^2$.
   (b) Use (a) to show that the infinite series converges absolutely, because the limit of the partial sums $\sum_{i=1}^{n} |x_i y_i|$ exists.
   (c) $e - 1$.
   (d) Note that $[1/i(i + 1)] = (1/i) - [1/(i + 1)]$. Hence the partial sum $\sum_{i=1}^{n} [1/i(i + 1)] = 1 - [1/(n + 1)]$ approaches 1, and $(\alpha, \beta) = 1$.

10. $(\alpha, \beta) = 17 - 7i$. $(\beta, \alpha) = 17 + 7i$. $\|\alpha\| = \sqrt{15}$. $\|\beta\| = 5$. $|(\alpha, \beta)| = \sqrt{338} \leq 5\sqrt{15}$.

11. $(\beta + \gamma, \alpha) = \overline{(\alpha, \beta + \gamma)} = \ldots = (\beta, \alpha) + (\gamma, \alpha)$.

12. $(a\alpha, \beta) = \overline{(\beta, a\alpha)} = \overline{a}\overline{(\beta, \alpha)} = a(\alpha, \beta)$. So we may use $(a\alpha, b\beta) = a\bar{b}(\alpha, \beta)$ and the two distributive laws.

13. Start with the right member and use the two preceding exercises.

7

ORTHOGONALITY

You may recall that in three-space two geometric vectors are perpendicular if their dot product is zero. In this lesson we extend the concept of perpendicularity (orthogonality) to inner product spaces of dimension higher than three. We will use the adjective "orthogonal" in several different ways, so you should be careful lest you misinterpret its meaning.

7-1 Definition

Two vectors α, β in an inner product space V are said to be *orthogonal* if their inner product is 0. The subset $A \subset V$ is said to be an *orthogonal* set if each pair of distinct vectors in A is orthogonal. The two subsets A and B are said to be *mutually orthogonal* if for all $\alpha \in A$ and $\beta \in B$, the inner product $(\alpha, \beta) = 0$. An *orthonormal* set is an orthogonal set of unit vectors.

Example 1. Let $V = R^4$ with the usual dot product. The vectors $(1, 2, 3, 4)$ and $(-9, 1, 1, 1)$ are orthogonal, because their dot product is zero. The set $A = \{(1, 1, 1, 0), (1, 0, -1, 0), (1, -2, 1, 3)\}$ is an orthogonal set, because the dot product of each pair is zero. The set A can be changed into an orthonormal set by multiplying each vector by the reciprocal of its length. This gives

$$\left\{ \frac{1}{\sqrt{3}}(1, 1, 1, 0), \frac{1}{\sqrt{2}}(1, 0, -1, 0), \frac{1}{\sqrt{15}}(1, -2, 1, 3) \right\}$$

(The above process of changing a vector into a unit vector is called "normalizing" the vector.)

Let $B = \{(1, 1, 0, 0), (1, 0, 1, 0)\}$ and $C = \{(1, -1, -1, 0), (0, 0, 0, 1), (1, -1, -1, 2)\}$. Then B and C are mutually orthogonal because the

Orthogonality

inner product of any two vectors, one from B and the other from C, is zero.

Example 2. Let $V = C(-1, 1)$, the space of functions continuous on $[-1, 1]$, with the usual inner product $(f, g) = \int_{-1}^{1} f(x)g(x)\,dx$. The set of functions $\{1, x, 3x^2 - 1\}$ is orthogonal, because the inner product of each pair is zero. (Check this.) The lengths of the three vectors are respectively $\sqrt{2}, \sqrt{2/3}, \sqrt{8/5}$. Therefore normalizing the set gives the orthonormal set

$$\left\{\frac{1}{\sqrt{2}}, \sqrt{\frac{3}{2}}x, \sqrt{\frac{5}{8}}(3x^2 - 1)\right\}$$

These are the first three so-called *Legendre polynomials* (normalized), of great importance in applied differential equations.*

Example 3. Let $V = C(0, 2\pi)$ with the usual inner product, and let A be the infinite subset $\{1, \sin x, \cos x, \sin 2x, \cos 2x, \ldots\}$ of functions from V. This is a very useful subset of A because its linear combinations are used extensively in successfully approximating periodic functions such as the functions which appear in electrical engineering and physics. An infinite series $a_0 + a_1 \sin x + b_1 \cos x + a_2 \sin 2x + b_2 \cos 2x + \ldots$ is called a "Fourier" series.[†]

We show here that the set A is orthogonal. There are only three inner products to evaluate. For these we use in turn the three identities

$$\sin a \sin b = \frac{1}{2}[\cos(a - b) - \cos(a + b)]$$

$$\cos a \cos b = \frac{1}{2}[\cos(a + b) + \cos(a - b)]$$

$$\sin a \cos b = \frac{1}{2}[\sin(a + b) + \sin(a - b)]$$

---

*The French mathematician Adrien Marie Legendre (1752–1833) is remembered for his work in integral calculus, even though his decades of efforts in elliptic integrals were nullified by simpler works of Abel and Jacobi. Legendre and Gauss share credit for the method of least squares. It was unfortunate that Legendre became bitter toward Gauss over this point.

†Joseph Fourier (1768–1830), a French mathematician, made many contributions to mathematical physics. He is best known for his Fourier series and its applications to physics.

For $m \geq 1$, $n \geq 1$, $m \neq n$, we have

$$(\sin mx, \sin nx) = \int_0^{2\pi} \sin mx \sin nx \, dx$$

$$= \frac{1}{2} \int_0^{2\pi} [\cos(m-n)x - \cos(m+n)x] \, dx = 0$$

For $m \geq 0$, $n \geq 0$, $m \neq n$,

$$(\cos mx, \cos nx) = \frac{1}{2} \int_0^{2\pi} [\cos(m+n)x + \cos(m-n)x] \, dx = 0$$

Finally, for $m \geq 1$, $n \geq 0$, $m \neq n$,

$$(\sin mx, \cos nx) = \frac{1}{2} \int_0^{2\pi} [\sin(m+n)x + \sin(m-n)x] \, dx = 0$$

It will be left as an exercise for the student to normalize the set.

Orthogonal sets and orthonormal sets are useful in applied mathematics, appearing in both n-tuple spaces and in function spaces. As we develop further properties, you should try to visualize each new result in three-space *without permitting this to beguile you from considering the problem properly in spaces of higher dimension.*

By geometric arguments we see that three mutually orthogonal nonzero vectors in R^3 form a linearly independent set. This result can be extended.

7-2 Theorem

In a real inner product space, any orthogonal set of nonzero vectors is linearly independent.

Proof: Let $\sum_{i=1}^{n} a_i \alpha_i = 0$ be a linear relation of the orthogonal set A. Then pick *any* particular integer k from the set $\{1, 2, \ldots, n\}$. The inner product

$$\left(\alpha_k, \sum_{i=1}^{n} a_i \alpha_i\right) = \sum_{i=1}^{n} a_i (\alpha_k, \alpha_i)$$

by the rules governing inner products. But the summation on the right contains at most one nonzero term, namely $a_k(\alpha_k, \alpha_k)$, because α_k and α_i are orthogonal by hypothesis when $k \neq i$. This gives $a_k = 0$ (for $k = 1, 2, \ldots, n$) and the linear relation we started with is trivial. Therefore, the set A is linearly independent. ∎

As a result we do not have to check an orthogonal set of nonzero vectors for linear independence. Any such set is linearly independent.

7-3 Corollary

Any orthogonal set of n nonzero vectors in an n-dimensional linear space V is a basis of V.

Proof: We have already proved that in n-space any linearly independent set of n vectors is a basis. ∎

In two-space or three-space, let α and β be two nonzero vectors. By the *projection of α upon β* we mean the vector obtained by projecting orthogonally the tip of α onto the line containing β. (See Figure 7-1.)

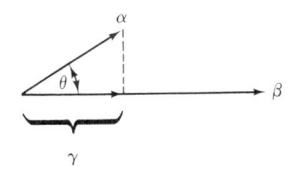

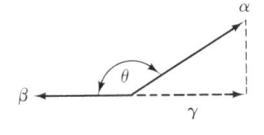

FIGURE 7-1

By elementary trigonometry γ has length $\|\alpha\| \cos \theta$ or $(\alpha \cdot \beta)/\|\beta\|$. We produce γ by taking the proper multiple of the unit vector $(1/\|\beta\|)\beta$. That is,

$$\gamma = \|\gamma\|\frac{1}{\|\beta\|}\beta = \frac{\alpha \cdot \beta}{\beta \cdot \beta}\beta$$

We extend the notion to a general Euclidean space.

7-4 Definition

Let α and β be vectors in a Euclidean space, with $\beta \neq 0$. The *projection of α upon β* is defined as the vector $[(\alpha, \beta)/(\beta, \beta)]\beta$.

Example 4. In R^4, let $\alpha = (1, 1, 1, 1)$, $\beta = (1, 0, 2, 0)$, with the usual dot product. The projection of α upon β is $(3/5)(1, 0, 2, 0)$, and the projection of β upon α is $(3/4)(1, 1, 1, 1)$.

Example 5. In $C(0, 2\pi)$ with the usual inner product $(f, g) = \int_0^{2\pi} f(x)g(x)\,dx$, the projection of x upon $\sin x$ is $[(x, \sin x)/(\sin x, \sin x)]\sin x$. Now

Lessons in Linear Algebra

$$(x, \sin x) = \int_0^{2\pi} x \sin x \, dx = \sin x - x \cos x \Big|_0^{2\pi} = -2\pi$$

$$(\sin x, \sin x) = \int_0^{2\pi} \sin^2 x \, dx = \frac{x}{2} - \frac{1}{4} \sin 2x \Big|_0^{2\pi} = \pi$$

so the projection of x upon $\sin x$ is $-2 \sin x$.

We shall observe next that $-2 \sin x$ is the multiple of $\sin x$ which is nearest x in the sense that $\|a \sin x - x\|$ is a minimum when $a = -2$. It will please you that the general proof for all Euclidean spaces is easier than checking out Example 5.

7-5 Theorem Let α and β ($\beta \neq 0$) be two vectors in the real inner product space V. Then the projection of α upon β is the multiple of β which is nearest α in the sense that $\|t\beta - \alpha\|$ is a minimum.

Proof: $\|t\beta - \alpha\|^2 = t^2 \|\beta\|^2 - 2t(\alpha, \beta) + \|\alpha\|^2$ is a quadratic expression in t, with the positive leading coefficient $\|\beta\|^2$. Therefore, the minimum is attained when $t = [2(\alpha, \beta)/2\|\beta\|^2]$ or when $t = (\alpha, \beta)/(\beta, \beta)$. ∎

In ordinary Cartesian geometry, the coordinates of a point are the lengths of the projections of the point on the axes. There is a similar fact in n-dimensional Euclidean spaces.

7-6 Theorem Let $A = \{\alpha_1, \alpha_2, \ldots, \alpha_n\}$ be an orthonormal basis of the Euclidean space V, and let β be any vector of V. Then $\beta = \sum_{i=1}^{n} a_i \alpha_i$, where, for each k ($k = 1, \ldots, n$), $a_k = (\beta, \alpha_k)$, the projection of β upon α_k.

Proof: $(\beta, \alpha_k) = (\sum_{i=1}^{n} a_i \alpha_i, \alpha_k) = \sum_{i=1}^{n} a_i(\alpha_i, \alpha_k)$, by the rules governing inner products. In the last summation the only nonzero term appears when i takes on the value k. Thus, $(\beta, \alpha_k) = a_k(\alpha_k, \alpha_k) = a_k$, because α_k is a unit vector. ∎

We conclude this lesson by showing that every finite-dimensional real inner product space has an orthonormal basis. Our proof will be "constructive" in that it gives a systematic method of changing a basis into an orthonormal basis. The real problem is to construct an orthogonal basis. Once we have done this, it is easy to normalize each vector.

Orthogonality

The idea has its roots in geometry. Let $\{\alpha_1, \alpha_2\}$ be a basis of R^2. We propose to construct an orthogonal basis by replacing α_2. (See Figure 7-2.) Let $\gamma = [(\alpha_2, \alpha_1)/(\alpha_1, \alpha_1)]\alpha_1$, the projection of α_2 upon α_1. Then $\alpha_2 - \gamma$ is orthogonal to α_1, and hence,

$$\left\{\alpha_1, \alpha_2 - \frac{(\alpha_2, \alpha_1)}{(\alpha_1, \alpha_1)}\alpha_1\right\}$$

is an orthogonal basis. The generalization follows.

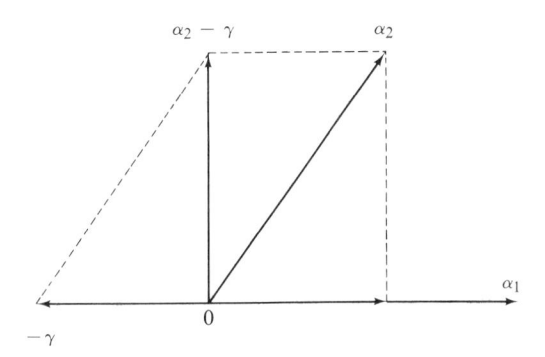

FIGURE 7-2

7-7 Theorem (*The Gram-Schmidt orthogonalization process.*) Let $A = \{\alpha_1, \alpha_2, \ldots, \alpha_n\}$ be a basis of the Euclidean space V, and let

$$\beta_1 = \alpha_1$$
$$\beta_2 = \alpha_2 - \frac{(\alpha_2, \beta_1)}{(\beta_1, \beta_1)}\beta_1$$
$$\beta_3 = \alpha_3 - \frac{(\alpha_3, \beta_2)}{(\beta_2, \beta_2)}\beta_2 - \frac{(\alpha_3, \beta_1)}{(\beta_1, \beta_1)}\beta_1$$
$$\vdots$$
$$\beta_n = \alpha_n - \frac{(\alpha_n, \beta_{n-1})}{(\beta_{n-1}, \beta_{n-1})}\beta_{n-1} - \cdots - \frac{(\alpha_n, \beta_1)}{(\beta_1, \beta_1)}\beta_1$$

Then $B = \{\beta_1, \ldots, \beta_n\}$ is an orthogonal basis of V.

Proof: The space $L(A)$ spanned by A is a subspace of $L(B)$, because each α_i is in $L(B)$ by the above equations. This means B is a basis because it is a spanning set with exactly n elements. It remains to show that B is orthogonal.

$$(\beta_1, \beta_2) = \left(\alpha_1, \alpha_2 - \frac{(\alpha_2, \beta_1)}{(\beta_1, \beta_1)}\beta_1\right)$$

$$= (\alpha_1, \alpha_2) - \frac{(\alpha_2, \beta_1)}{(\beta_1, \beta_1)}(\alpha_1, \beta_1) \quad \text{(distributivity)}$$

$$= (\alpha_1, \alpha_2) - (\alpha_2, \alpha_1) = 0 \quad \text{(remember } \alpha_1 = \beta_1)$$

Now we use mathematical induction. Let k be any positive integer less than n for which $\{\beta_1, \beta_2, \ldots, \beta_k\}$ is orthogonal. We use this induction hypothesis and show that $\{\beta_1, \ldots, \beta_{k+1}\}$ is orthogonal. For each i in the set $1, 2, \ldots, k$,

$$(\beta_{k+1}, \beta_i) = \left(\alpha_{k+1} - \frac{(\alpha_{k+1}, \beta_k)}{(\beta_k, \beta_k)}\beta_k - \cdots - \frac{(\alpha_{k+1}, \beta_1)}{(\beta_1, \beta_1)}\beta_1, \beta_i\right)$$

$$= (\alpha_{k+1}, \beta_i) - \frac{(\alpha_{k+1}, \beta_i)}{(\beta_i, \beta_i)}(\beta_i, \beta_i) = 0$$

Most of the terms dropped out because β_i is orthogonal to each other element of $\{\beta_1, \ldots, \beta_k\}$. ∎

Example 6. We use the Gram-Schmidt process to construct an orthonormal basis of the three-dimensional subspace of R^4 spanned by the basis $A = \{(1, 1, 0, 0), (1, -2, 1, 0), (0, 1, 1, 1)\}$.

Let $\alpha_1, \alpha_2, \alpha_3$ be the vectors of A, and construct $\beta_1, \beta_2, \beta_3$.

$\beta_1 = \alpha_1 = (1, 1, 0, 0)$

$\beta_2 = \alpha_2 - \dfrac{(\alpha_2, \beta_1)}{(\beta_1, \beta_1)}\beta_1 = (1, -2, 1, 0) - \dfrac{-1}{2}(1, 1, 0, 0) = \left(\dfrac{3}{2}, -\dfrac{3}{2}, 1, 0\right)$

$\beta_3 = \alpha_3 - \dfrac{(\alpha_3, \beta_2)}{(\beta_2, \beta_2)}\beta_2 - \dfrac{(\alpha_3, \beta_1)}{(\beta_1, \beta_1)}\beta_1$

$= (0, 1, 1, 1) - \dfrac{-\frac{1}{2}}{\frac{11}{2}}\left(\dfrac{3}{2}, -\dfrac{3}{2}, 1, 0\right) - \dfrac{1}{2}(1, 1, 0, 0)$

$= \left(-\dfrac{4}{11}, \dfrac{4}{11}, \dfrac{12}{11}, 1\right)$

Therefore, $\{(1, 1, 0, 0), (3, -3, 2, 0), (-4, 4, 12, 11)\}$ is an orthogonal basis, and $\{1/\sqrt{2}\,(1, 1, 0, 0), 1/\sqrt{22}\,(3, -3, 2, 0), 1/\sqrt{297}\,(-4, 4, 12, 11)\}$ is an orthonormal basis.

The drudgery of the work could have been reduced by replacing the β's, as they appear, with attractive multiples before going on to the the next step. For example, we could have replaced β_2 with $(3, -3, 2, 0)$ before computing β_3.

Actually, the Gram-Schmidt process does more than is indicated by Theorem 7-7. In case we start with a linearly dependent set A, then applying the Gram-Schmidt process gives an orthogonal set which spans $L(A)$.

7-8 Corollary

In an n-dimensional Euclidean space, each orthogonal (orthonormal) set of nonzero vectors is part of an orthogonal (orthonormal) basis.

Proof: Let $\{\alpha_1, \ldots, \alpha_k\}$ be orthogonal and let $A = \{\alpha_1, \ldots, \alpha_n\}$ be a basis. Applying the Gram-Schmidt process to A gives the orthogonal basis $\{\alpha_1, \alpha_2, \ldots, \alpha_k, \beta_{k+1}, \ldots, \beta_n\}$, because the first k vectors are unchanged by the process. ∎

Optional Section on Complex Spaces

Most of the results in the lesson can be extended into complex inner product spaces. Theorem 7-2 uses the fact that $(a\alpha, b\beta) = a\bar{b}(\alpha, \beta)$.

The analog of Theorem 7-5 follows, with a proof which also serves as an alternate proof to Theorem 7-5.

7-9 Theorem

Let α and β ($\beta \neq 0$) be two vectors in a complex inner product space. Then the projection of α upon β is the multiple of β which is nearest to α.

Proof: Let $a = (\alpha, \beta)/(\beta, \beta)$ so that $a\beta$ is the projection of α upon β. We will compare $\|a\beta - \alpha\|$ with $\|t\beta - \alpha\|$, where t is an arbitrary scalar, and show that $\|a\beta - \alpha\|$ is the desired minimum. Observe that $t\beta - \alpha = (t\beta - a\beta) + (a\beta - \alpha) = (t - a)\beta + (a\beta - \alpha)$ and hence,

$$\|t\beta - \alpha\|^2 = \|(t - a)\beta\|^2 + ((t - a)\beta, a\beta - \alpha)$$
$$+ (a\beta - \alpha, (t - a)\beta) + \|a\beta - \alpha\|^2$$

The two middle terms of the right member drop out, because $(a\beta - \alpha, \beta) = a(\beta, \beta) - (\alpha, \beta) = 0$. We then have

$$\|t\beta - \alpha\|^2 = \|(t - a)\beta\|^2 + \|a\beta - \alpha\|^2$$

which attains its minimum when $t = a$. ∎

Lessons in Linear Algebra

The remaining theorems for complex spaces are mentioned briefly. Theorems 7-6 and 7-7 hold for complex inner product spaces, and the proofs are almost identical with the proofs for real spaces. You must, however, remember that the scalars are complex numbers and that laws such as $(a\alpha, b\beta) = a\bar{b}(\alpha, \beta)$ prevail.

Summary of Topics in Lesson 7

Orthogonality in Euclidean spaces.
Two orthogonal vectors.
An orthogonal set of vectors.
An orthonormal set of vectors.
Two mutually orthogonal sets.
Linear independence assured for an orthogonal set.
Orthogonal and orthonormal bases.
Projection of one vector upon another.
The projection of α upon β is the vector of the set $\{a\beta\}$ which is nearest α.
Construction of an orthogonal basis by the Gram-Schmidt process.
Optional results in complex spaces.

Exercises

1. Find two nonzero vectors in R^4, orthogonal to each other and to the vector $(2, 1, -3, 1)$.

2. Find a unit vector orthogonal to each of the vectors $(2, -1, 5, 4)$ and $(3, 1, 1, 0)$.

3. Find an orthonormal basis of R^3 such that one vector of the basis is a multiple of $(2, 2, 1)$.

4. In studying the methane molecule it is helpful to know the angle made by two lines drawn from the center of a regular tetrahedron to a pair of vertices. Find the angle by inscribing the tetrahedron in a cube whose center is at the origin and two of whose vertices are $(1, 1, 1)$ and $(1, -1, -1)$. Recall the formula $\alpha \cdot \beta = \|\alpha\| \|\beta\| \cos \theta$.

5. Electrical engineers often consider functions similar to the one whose graph in the xy-plane is a sawtooth wave of period π, such that $y = x$ on the interval $[0, \pi/2]$, and $y = \pi - x$ on the interval $[\pi/2, \pi]$. (See Figure 7-3). Find the multiple of $\sin x$ which, on the

Orthogonality

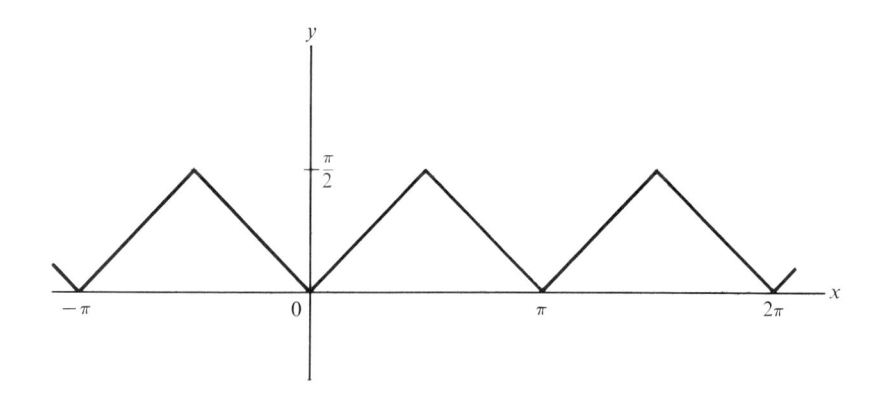

FIGURE 7-3

interval $[0, \pi]$, is nearest the function, and sketch the graph on the interval $[0, \pi]$. Use $(f, g) = \int_0^\pi f(x)g(x)\, dx$.

6. A social scientist wishes to approximate the function e^x on the interval $[0, 1]$. Use $(f, g) = \int_0^1 f(x)g(x)\, dx$.
   (a) Find a such that the constant function a is nearest e^x. (Find the projection of e^x upon 1.)
   (b) Find a such that ax is nearest e^x.
   (c) Find a such that ax^2 is nearest e^x.

7. Prove that if S and T are mutually orthogonal subspaces, then $S \cap T = \{0\}$.

8. Apply the Gram-Schmidt process to the basis $\alpha_1 = (1, -1, 2)$, $\alpha_2 = (1, 1, 1)$, $\alpha_3 = (2, -1, 1)$ of R^3.

9. Find an orthonormal basis for the space spanned by $\{(1, 1, 2, 1), (-1, 1, 0, 1), (2, 1, 1, 0)\}$.

10. Find an orthonormal basis for the space spanned by $\{(1, 1, 2, 1), (-1, 1, 0, 1), (-1, 3, 2, 3)\}$.

11. In $C(-1, 1)$, with $(f, g) = \int_{-1}^1 f(x)g(x)\, dx$, apply the Gram-Schmidt process to the basis $\{1, x, x^2\}$ of the subspace V in order to obtain an orthonormal basis of V.

12. In $C(0, 2\pi)$ convert the orthogonal set $\{1, \sin x, \cos x, \ldots, \sin nx, \cos nx, \ldots\}$ into an orthonormal set.

79

Lessons in Linear Algebra

13. Let $\{\alpha_1, \ldots, \alpha_n\}$ be an orthonormal basis of a Euclidean space, and let β, γ be two vectors with $\beta = \sum_{i=1}^{n} b_i\alpha_i$, $\gamma = \sum_{i=1}^{n} c_i\alpha_i$. Show that $(\beta, \gamma) = \sum_{i=1}^{n} b_i c_i$. This shows that any inner product is a dot product relative to an orthonormal basis.

Optional Exercises on Complex Spaces

In Exercises 14, 15, 17, and 20, take for $\alpha = (a_1, a_2, \ldots)$ and $\beta = (b_1, b_2, \ldots)$ the inner product $(\alpha, \beta) = a_1\bar{b}_1 + a_2\bar{b}_2 + \ldots$.

14. Find a unit vector in C^3, orthogonal to each of the vectors $(1, 2, 3 + i), (-1 + i, i, 2)$.

15. Find an orthonormal basis of C^2 such that one vector of the basis is a multiple of $(2 + 2i, i)$.

16. Prove that in a complex inner product space, any orthogonal set of nonzero vectors is linearly independent. (Parallel the proof of Theorem 7-2.)

17. In C^2, find the projection of $(1 + i, 2 - i)$ upon $(3, 4 - i)$.

18. Prove Theorem 7-6 for a complex inner product space V. Let $A = \{\alpha_1, \ldots, \alpha_n\}$ be an orthonormal basis of V and let $\beta \in V$. Then show $\beta = \sum_{i=1}^{n} (\beta, \alpha_i)\alpha_i$.

19. Prove that the Gram-Schmidt orthogonalization process works for a complex inner product space of finite dimension. (Parallel the proof of Theorem 7-7.)

20. Use the Gram-Schmidt process to construct an orthonormal basis for the space spanned by $\{(i, 2, 2i), (1 + 2i, 0, 2)\}$.

21. Prove for complex inner product spaces a result analogous to Exercise 13.

Remarks on Exercises

1. The answer is not unique. One possiblity is the pair $\{(0, 1, 0, -1), (3, 0, 2, 0)\}$, obtained by inspection.

2. Find (a, b, c, d) such that $2a - b + 5c + 4d = 0$, $3a + b + c = 0$. One solution is $(-2, 5, 1, 1)$. This leads to the unit vector

$$\left(\frac{1}{\sqrt{31}}\right)(-2, 5, 1, 1)$$

Orthogonality

3. One orthogonal basis is $\{(2, 2, 1), (1, -2, 2), (2, -1, -2)\}$. Normalizing gives the orthonormal basis
$$\left\{\frac{1}{3}(2, 2, 1), \frac{1}{3}(1, -2, 2), \frac{1}{3}(2, -1, -2)\right\}$$
The answer is not unique

4. For $\alpha = (1, 1, 1)$ and $\beta = (1, -1, -1)$, $\cos\theta = (\alpha \cdot \beta)/\|\alpha\| \|\beta\| = -1/3$.

5. The required function is the projection of f upon $\sin x$,
$$\frac{(f, \sin x)}{(\sin x, \sin x)} \sin x = \frac{2}{\frac{\pi}{2}} \sin x = \frac{4}{\pi} \sin x$$

6. (a) $e - 1$.
   (b) $a = 3$ gives the function $3x$.
   (c) $a = 5(e - 2)$ gives the function $5(e - 2)x^2$.

8. $\left\{\dfrac{1}{\sqrt{6}}(1, -1, 2), \dfrac{1}{\sqrt{21}}(2, 4, 1), \dfrac{1}{\sqrt{14}}(3, -1, -2)\right\}$.

9. $\left\{\dfrac{1}{\sqrt{7}}(1, 1, 2, 1), \dfrac{1}{\sqrt{35}}(-4, 3, -1, 3), \dfrac{1}{\sqrt{35}}(3, 4, -3, -1)\right\}$.

10. $\left\{\dfrac{1}{\sqrt{7}}(1, 1, 2, 1), \dfrac{1}{\sqrt{35}}(-4, 3, -1, 3)\right\}$.

11. $\left\{\dfrac{1}{\sqrt{2}}, \sqrt{\dfrac{3}{2}}x, \sqrt{\dfrac{5}{8}}(3x^2 - 1)\right\}$. Also, see Example 2.

12. $\left\{\dfrac{1}{\sqrt{2\pi}}, \dfrac{1}{\sqrt{\pi}}\sin x, \dfrac{1}{\sqrt{\pi}}\cos x, \ldots, \dfrac{1}{\sqrt{\pi}}\sin nx, \dfrac{1}{\sqrt{\pi}}\cos nx, \ldots\right\}$.

13. $(\beta, \gamma) = \left(\sum_{i=1}^{n} b_i\alpha_i, \sum_{j=1}^{n} c_j\alpha_j\right) = \sum_i \sum_j b_i c_j (\alpha_i, \alpha_j)$
$= \sum_i b_i c_i (\alpha_i, \alpha_i) = \sum_i b_i c_i.$

14. $\pm \dfrac{1}{\sqrt{79}}(5 + 3i, -6 - 2i, 2 + i)$.

15. $\left\{\dfrac{1}{3}(2 + 2i, i), \dfrac{1}{3}(i, 2 - 2i)\right\}$.

17. $\dfrac{12 + i}{26}(3, 4 - i)$ or $\dfrac{1}{26}(36 + 3i, 49 - 8i)$.

20. $\left\{\dfrac{1}{3}(i, 2, 2i), \dfrac{1}{\sqrt{117}}(2 + 8i, -2 + 5i, 4 - 2i)\right\}$.

8

ORTHOGONAL COMPLEMENTS

In this lesson we continue our study of orthogonality. We define the *orthogonal complement* of a subset of vectors and we develop a method for "projecting" a vector α upon a subspace S in order to obtain the vector in S nearest to α. The whole idea is revealed by geometric preliminaries.

Let α be a vector in ordinary 3-space, and let S be a plane through the origin. (See Figure 8-1.) As usual our vectors in this little example are

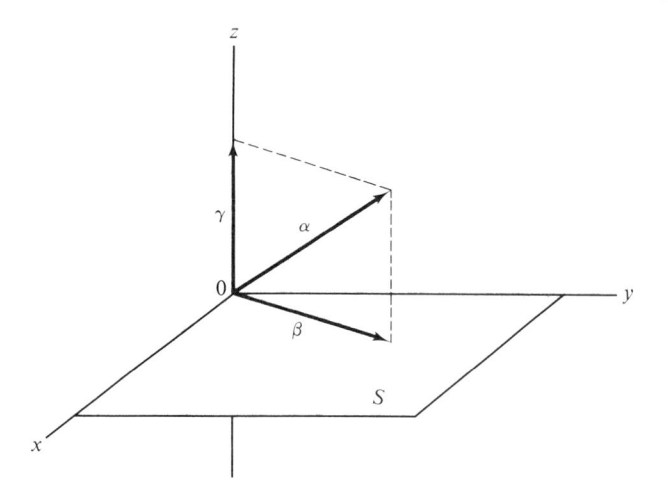

FIGURE 8-1

directed line segments with initial points at the origin. The "orthogonal complement" of S is the set of all vectors orthogonal to S, in this case the z-axis, a one-dimensional subspace. Observe that α can be expressed uniquely as the sum of two vectors β and γ, one in S and one in the orthogonal complement of S. Furthermore, β is the vector in S which is nearest to α, and γ is the vector in the orthogonal complement which is nearest to α.

Also, the Pythagorean theorem tells us that $\|\alpha\|^2 = \|\beta\|^2 + \|\gamma\|^2$.

All of the above can be extended into more general spaces. We give now the details for real inner product spaces (Euclidean spaces).

8-1 Definition

Let S be a subset of a Euclidean space V. The *orthogonal complement of* S, denoted $S^\perp$, is the set of all vectors orthogonal to S. That is

$$S^\perp = \{\alpha \in V \mid (\alpha, \sigma) = 0 \text{ for all } \sigma \in S\}$$

(We will read $S^\perp$ as "S perpendicular" or as "S perp.")

The orthogonal complement is well-defined whether or not S is a subspace. However, $S^\perp$ itself is always a subspace. For example, in R^3, if $S = \{(1, 0, 0), (0, 1, 0)\}$, then $S^\perp = \{(0, 0, z)\}$, the z-axis. As another example, if $S = \{(x, y, 0)\}$, the subspace of vectors in the xy-plane, then $S^\perp$ is again the z-axis.

8-2 Theorem

Let S be a finite-dimensional subspace of the Euclidean space V. Then

(1) $S^\perp$ is a subspace.
(2) $S \cap S^\perp = \{0\}$ and $S + S^\perp = V$.
(3) if V is finite-dimensional, then $\dim S + \dim S^\perp = \dim V$ and $S^{\perp\perp} = S$.

Proof: (1) We show $S^\perp$ is closed under linear combinations. Let $\alpha, \beta \in S^\perp$ and let $\sigma \in S$. Then $(a\alpha + b\beta, \sigma) = a(\alpha, \sigma) + b(\beta, \sigma) = 0 + 0 = 0$, and hence, $a\alpha + b\beta$ is in $S^\perp$.

(2) $S \cap S^\perp = \{0\}$ because the only self-orthogonal vector is 0. Let $\{\alpha_1, \ldots, \alpha_m\}$ be an orthonormal basis of S, and let $\beta \in V$. To show that $S + S^\perp = V$ we show that β is the sum of two vectors σ and τ, one in S and the other in $S^\perp$.

Take $\sigma = \sum_{i=1}^{m} (\beta, \alpha_i) \alpha_i$, clearly a vector in S, and define τ by the equation $\beta = \sigma + \tau$. Then τ is in $S^\perp$, because for $k = 1, 2, \ldots, m$ we have

$$(\tau, \alpha_k) = (\beta - \sigma, \alpha_k)$$
$$= (\beta, \alpha_k) - \left(\sum_{i=1}^{m} (\beta, \alpha_i)\alpha_i, \alpha_k\right)$$
$$= (\beta, \alpha_k) - (\beta, \alpha_k) = 0$$

(Observe that β is the sum of the projections of β upon the vectors in the orthonormal basis of S.)

(3) Let dim $V =$ dim$(S + S^\perp) = n$. By Theorem 4-4, dim $(S + S^\perp) =$ dim $S +$ dim $S^\perp -$ dim$(S \cap S^\perp)$. Then using $S \cap S^\perp = \{0\}$ gives $n =$ dim $S +$ dim$(S^\perp)$.

Finally, we show that $S^{\perp\perp} = S$. By the definition of orthogonal complement, $S \subset S^{\perp\perp}$. Take dim $S = m$. Then dim $S^\perp = n - m$ and dim $S^{\perp\perp} = n - (n - m) = m$. Having shown that $S \subset S^{\perp\perp}$ and dim $S =$ dim $S^{\perp\perp}$, we conclude that $S = S^{\perp\perp}$. ∎

Example 1. In R^5, let $A = \{(1, 0, 0, 0, 0), (0, 1, 0, 0, 0)\}$, $B = \{(0, 0, 1, 0, 0), (0, 0, 0, 1, 0), (0, 0, 0, 0, 1)\}$. Then $L(A)$ and $L(B)$ are orthogonal complements of each other.

Example 2. In $C(0, 2\pi)$, let V be the space spanned by $\{1, \sin x, \cos x, \sin 2x, \cos 2x\}$. Then the space spanned by $\{1, \cos x, \cos 2x\}$ is the orthogonal complement of the space spanned by $\{\sin x, \sin 2x\}$. We showed in Lesson 7, Example 3, that the set $\{1, \sin x, \cos x, \sin 2x, \cos 2x\}$ is orthogonal under the usual inner product $(f, g) = \int_0^{2\pi} f(x)g(x)\, dx$.

8-3 Theorem

Let V be a Euclidean space, and let S be a finite-dimensional subspace of V. Then each vector β of V can be expressed uniquely as $\beta = \sigma + \tau$, the sum of two vectors, one in S and the other in $S^\perp$.

Proof: We showed in Theorem 8-2 (part 2) that $V = S + S^\perp$, which means that each vector of V is the sum of two vectors, one from S and the other from $S^\perp$. We have only to show uniqueness.

Suppose there were two such representations, $\beta = \sigma_1 + \tau_1 = \sigma_2 + \tau_2$ where $\sigma_1, \sigma_2 \in S$ and $\tau_1, \tau_2 \in S^\perp$. The vector $\sigma_1 - \sigma_2 = \tau_2 - \tau_1$. But the only vector in $S \cap S^\perp$ is 0, and we have $\sigma_1 - \sigma_2 = \tau_2 - \tau_1 = 0$. ∎

Example 3. Let $S = \{(x, y, y, 0) \mid x, y \in R\}$, a subspace of R^4 of dimension 2. A basis (orthogonal but not orthonormal) of S is $\{(1, 0, 0, 0), (0, 1, 1, 0)\}$. A basis of the orthogonal complement T is $\{(0, 1, -1, 0), (0, 0, 0, 1)\}$. The vector $\beta = (1, 3, 1, 3)$ can be expressed in just one way as the sum of two vectors $\sigma \in S$ and $\tau \in T$, namely with $\sigma = (1, 2, 2, 0)$ and $\tau = (0, 1, -1, 3)$. This illustrates Theorem 8-3.

Orthogonal Complements

Observe also that $\|\beta\|^2 = \|\sigma\|^2 + \|\tau\|^2$, because $20 = 9 + 11$. The generalization will be Theorem 8.4.

In Exercise 13 of Lesson 7, it is pointed out that the inner product is easy to handle relative to an orthonormal basis. Specifically, if β and γ are given in terms of an orthonormal basis $\{\alpha_1, \ldots, \alpha_n\}$ by $\beta = \sum_{i=1}^{n} b_i \alpha_i$ and $\gamma = \sum_{i=1}^{n} c_i \alpha_i$, then $(\beta, \gamma) = \sum_{i=1}^{n} b_i c_i$.

Example 4. In R^2 with the usual inner product, let $\beta = (3, 1)$, $\gamma = (1, -5)$. Then $(\beta, \gamma) = -2$.

The vectors $\alpha_1 = (1/\sqrt{2})(1, 1)$, $\alpha_2 = (1/\sqrt{2})(1, -1)$ form an orthonormal basis, and

$$\beta = 2\sqrt{2}\alpha_1 + \sqrt{2}\alpha_2$$
$$\gamma = -2\sqrt{2}\alpha_1 + 3\sqrt{2}\alpha_2$$

Observe that $(2\sqrt{2}, \sqrt{2}) \cdot (-2\sqrt{2}, 3\sqrt{2}) = -2 = (\beta, \gamma)$.

Example 5. In $C(0, 2\pi)$ with the usual inner product, the set $\alpha_1 = (1/\sqrt{\pi}) \sin x$, $\alpha_2 = (1/\sqrt{\pi}) \cos x$ is orthonormal. Let $\beta = \sin x + \cos x$, $\gamma = 2 \sin x - 3 \cos x$. Then $\beta = \sqrt{\pi}\alpha_1 + \sqrt{\pi}\alpha_2$, $\gamma = 2\sqrt{\pi}\alpha_1 - 3\sqrt{\pi}\alpha_2$, and $(\beta, \gamma) = (\sqrt{\pi})(2\sqrt{\pi}) + (\sqrt{\pi})(-3\sqrt{\pi}) = -\pi$.

Next we prove a Pythagorean theorem.

8-4 Theorem Let S be a finite-dimensional subspace of the Euclidean space V, and let $\beta = \sigma + \tau$, where $\sigma \in S$ and $\tau \in S^\perp$. Then $\|\beta\|^2 = \|\sigma\|^2 + \|\tau\|^2$.

Proof:
$$(\beta, \beta) = (\sigma + \tau, \sigma + \tau)$$
$$= (\sigma, \sigma) + 2(\sigma, \tau) + (\tau, \tau)$$
$$= (\sigma, \sigma) + (\tau, \tau). \blacksquare$$

8-5 Definition Let S be a finite-dimensional subspace of the Euclidean space V, and let $\beta = \sigma + \tau$, where $\sigma \in S$, and $\tau \in S^\perp$. Then σ is called the *projection of β upon S*.

Lessons in Linear Algebra

Example 6. In $C(0, 2\pi)$, let S be the space spanned by the orthogonal set $\{\sin x, \cos x\}$ and let $\beta = x$. Then σ, the projection of β upon S, is the sum of the projections of β upon $\sin x$ and $\cos x$, the results being the same as if the basis were orthonormal. We have

$$\sigma = \frac{(x, \sin x)}{(\sin x, \sin x)} \sin x + \frac{(x, \cos x)}{(\cos x, \cos x)} \cos x$$

$$= \frac{-2\pi}{\pi} \sin x + \frac{0}{\pi} \cos x \quad \text{(See Lesson 7, Example 5.)}$$

$$= -2 \sin x$$

Then $\beta = \sigma + \tau$ or

$$x = \underbrace{(-2 \sin x)}_{\sigma} + \underbrace{(x + 2 \sin x)}_{\tau}$$

where $(\sigma, \tau) = 0$. (Check the integration.)
Also,

$$\|\sigma\|^2 = \int_0^{2\pi} 4 \sin^2 x \, dx = 4\pi$$

$$\|\tau\|^2 = \int_0^{2\pi} (x + 2 \sin x)^2 \, dx = \frac{8\pi^3}{3} - 4\pi$$

$$\|\beta\|^2 = \int_0^{2\pi} x^2 \, dx = \frac{8\pi^3}{3}$$

which exhibits the Pythagorean relation.

All of the results above would be the same if we normalized the set $\{\sin x, \cos x\}$ before starting.

We conclude this lesson by showing that the projection of a vector is a useful approximation of the vector.

8-6 Theorem

Let S be a finite-dimensional subspace of the Euclidean space V and let $\beta \in V$. The projection of β upon S is the vector in S which is nearest to β.

Proof: Let $\beta = \sigma_1 + \tau_1$ where σ_1 is the projection of β upon S, and let σ be a variable vector of S. Then

$$\beta - \sigma = (\beta - \sigma_1) + (\sigma_1 - \sigma)$$

where $\beta - \sigma_1$ is in $S^\perp$ and $\sigma_1 - \sigma$ is in S. By Theorem 8-4 we have

$$\|\beta - \sigma\|^2 = \|\beta - \sigma_1\|^2 + \|\sigma_1 - \sigma\|^2$$

and it is apparent that $\|\beta - \sigma\|$ is a minimum when $\sigma = \sigma_1$. ∎

Orthogonal Complements

Example 7. In $C(-1, 1)$ we find the linear combination of $\{1, x, 3x^2 - 1\}$ which is the best approximation of $x^{2/3}$. The set $\{1, x, 3x^2 - 1\}$ is orthogonal by Exercise 11 of Lesson 7.

We simply add the required projections and obtain

$$\frac{(x^{2/3}, 1)}{(1, 1)} (1) + \frac{(x^{2/3}, x)}{(x, x)} x + \frac{(x^{2/3}, 3x^2 - 1)}{(3x^2 - 1, 3x^2 - 1)} (3x^2 - 1)$$

where $(x^{2/3}, 1) = \int_{-1}^{1} x^{2/3} \, dx = (6/5)$, etc. This gives (after some computation) the function

$$\frac{3}{5}(1) + 0\,(x) + \frac{3}{11}(3x^2 - 1) \quad \text{or} \quad \frac{18}{55} + \frac{9}{11} x^2$$

We have a parabola (shown in Figure 8-2) which approximates on the interval $[-1, 1]$ the semicubical parabola $y = x^{2/3}$.

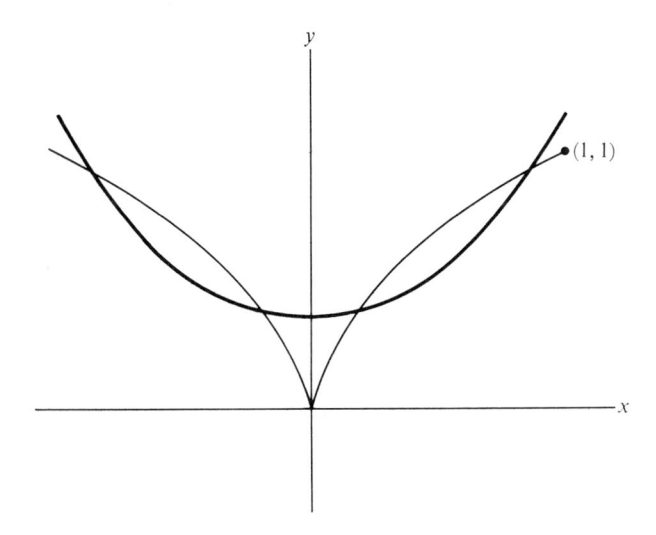

FIGURE 8-2

Optional Section on Complex Spaces

The results of this lesson hold with minor changes for complex inner product spaces. The *orthogonal complement* $S^\perp$ of a finite-dimensional subspace is defined the same way, by Definition 8-1, and *projection* means the same thing (Definition 8-5).

All of the Theorems 8-2, 8-3, 8-4, and 8-6 hold. Furthermore, the proofs given are valid, with practically no changes. The only change is a nonvital one in the proof of Theorem 8-4.

Example 8. In C^2, let S be the one-dimensional subspace spanned by $\alpha = (1, i)$ and let $\beta = (3i, 1)$. Using the inner product $[(a_1, a_2), (b_1, b_2)] = a_1\bar{b}_1 + a_2\bar{b}_2$, we have the projection of β upon S,

$$\sigma = \frac{(\beta, \alpha)}{(\alpha, \alpha)} \alpha = \frac{2i}{2}(1, i) = (i, -1)$$

Then $\beta = \sigma + \tau$, where $\tau = (2i, 2)$ and is in $S^\perp$, because $(\sigma, \tau) = 0$.

The Pythagorean relation $\|\beta\|^2 = \|\sigma\|^2 + \|\tau\|^2$ is verified by $10 = 2 + 8$.

Summary of Topics in Lesson 8

Orthogonal complement of a subset.

The orthogonal complement is a subspace.

A subspace of V with its orthogonal complement spans V.

The unique representation of a vector relative to a subspace and its orthogonal complement.

The projection of a vector upon a given subspace.

The vector in a subspace which is nearest to a given vector.

Optional results in complex spaces.

Exercises

1. In R^2, let S be the subspace spanned by the basis $\{(1, 0, 1), (0, 1, 1)\}$, and let $\beta = (5, 2, -2)$.
   (a) Find a basis of $S^\perp$.
   (b) Express $\beta = \sigma + \tau$, where $\sigma \varepsilon S$ and $\tau \varepsilon S^\perp$. Then check the Pythagorean relation $\|\beta\|^2 = \|\sigma\|^2 + \|\tau\|^2$. There are two approaches possible. You may orthogonalize the basis of S and find σ, or you may find τ first.
   (c) Also, β equals the sum of $\sigma_1 = (4, 1, 1)$ and $\tau_1 = (1, 1, -3)$. Check the Pythagorean relation $\|\beta\|^2 = \|\sigma_1\|^2 + \|\tau_1\|^2$.
2. In $C(0, \pi)$ let S be the space spanned by the orthogonal set $\{1, \cos x\}$ and let $\beta = x$. As usual $(f, g) = \int_0^\pi f(x)g(x)\,dx$.

(a) Express $\beta = \sigma + \tau$, where $\sigma \varepsilon S$ and $\tau \varepsilon S^{\perp}$.
(b) What function of S is nearest to x? Illustrate with a graph.

3. In R^5 let S be the subspace spanned by the orthogonal set $\{(1, 0, 1, 0, 1), (0, 1, 0, 1, 0)\}$, and let $\beta = (1, 3, 2, 1, 1)$.
   (a) Find the vector in S which is nearest to β.
   (b) Express β as $\sigma + \tau$, with $\sigma \varepsilon S$, $\tau \varepsilon S^{\perp}$, and check the Pythagorean relation.

4. Show the projection of α upon the nonzero vector β is equal to the projection of α upon any nonzero multiple of β.

5. In $C(0, 1)$, let S have basis $\{1, x\}$. Find the function of S which is nearest to $\sqrt{x}$, and graph. (You should first convert the basis into an orthogonal one.)

6. Let V be a Euclidean space.
   (a) Prove that $S \subset T$ implies $T^{\perp} \subset S^{\perp}$.
   (b) If S and T are subspaces, prove that $(S + T)^{\perp} = S^{\perp} \cap T^{\perp}$.
   (c) If V is finite dimensional, use (b) and the fact that $U^{\perp\perp} = U$ for all subspaces to show that $(S \cap T)^{\perp} = S^{\perp} + T^{\perp}$ for the subspaces S and T.

7. Suppose S and T are two subspaces of V whose direct sum is V. This means $S + T = V$ and $S \cap T = \{0\}$. Without assuming that V is an inner product space, prove that any vector β can be expressed uniquely as the sum of two vectors, one from S and the other from T.

8. The graph of an electric charge q is periodic of period 2π, and it is known that, on the interval $[0, 2\pi)$, $q(t) = 2\pi - t$. Find the linear combination $f(t)$ of $\{1, \sin t, \cos t\}$ which is nearest to $2\pi - t$ on the interval $[0, 2\pi]$. Then sketch $q(t)$ and $f(t)$ each from $x = 0$ to $x = 6\pi$.

9. Let $A = \{\alpha_1, \alpha_2, \ldots, \alpha_n\}$ be an orthonormal subset of V, and let $\beta \varepsilon V$. Prove that the "Bessel" inequality $\sum_{i=1}^{n} |\beta, \alpha_i|^2 \leq \|\beta\|^2$ holds.*

Optional Exercises on Complex Spaces

10. In C^3, let S be the subspace spanned by the orthogonal set $\{(1, 0, i), (1, 0, -i)\}$, and let $\beta = (1, i, 1)$. With the usual inner product,

---

*Friedrich Wilhelm Bessel (1784–1846), mathematician and astronomer, was a confidant of Gauss. Bessel's equations and functions are widely used in applied mathematics.

Lessons in Linear Algebra

     (a) find the vector σ in S nearest to β.
     (b) check the Pythagorean relation for $\beta = \sigma + \tau$.

11. Prove Theorem 8-4 for complex inner product spaces.

Remarks on Exercises

1. (a) $(1, 1, -1)$.
   (b) $\beta = \sigma + \tau = (2, -1, 1) + (3, 3, -3)$. $\|\sigma\|^2 + \|\tau\|^2 = 6 + 27 = 33 = \|\beta\|^2$.
   (c) $\|\sigma_1\|^2 + \|\tau_1\|^2 = 18 + 11 \neq \|\beta\|^2$.

2. (a) $\sigma = \dfrac{\pi}{2} - \dfrac{4}{\pi} \cos x$, $\tau = \beta - \sigma = x - \dfrac{\pi}{2} + \dfrac{4}{\pi} \cos x$.
   (b) σ. See Figure 8-3.

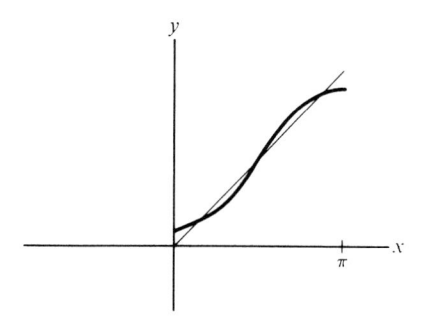

FIGURE 8-3

3. (a) $\sigma = \left(\dfrac{4}{3}, 2, \dfrac{4}{3}, 2, \dfrac{4}{3}\right)$.
   (b) σ from part (a). $\tau = \left(-\dfrac{1}{3}, 1, \dfrac{2}{3}, -1, -\dfrac{1}{3}\right)$. $\|\sigma\|^2 + \|\tau\|^2 = \dfrac{40}{3} + \dfrac{8}{3} = 16 = \|\beta\|^2$.

4. The projection of α upon $k\beta$ is
$$\dfrac{(\alpha, k\beta)}{(k\beta, k\beta)} k\beta = \dfrac{k(\alpha, \beta)}{k^2(\beta, \beta)} k\beta = \dfrac{(\alpha, \beta)}{(\beta, \beta)} \beta$$
the projection of α upon β.

5. Converting the basis $\{1, x\}$ to an orthogonal basis gives $\left\{1, x - \dfrac{1}{2}\right\}$.
The required function is (see Figure 8-4)

Orthogonal Complements

$$\frac{(\sqrt{x}, 1)}{(1, 1)}(1) + \frac{\left(\sqrt{x}, x - \frac{1}{2}\right)}{\left(x - \frac{1}{2}, x - \frac{1}{2}\right)}\left(x - \frac{1}{2}\right) = \frac{2}{3} + \frac{4}{5}\left(x - \frac{1}{2}\right) = \frac{4}{15} + \frac{4}{5}x$$

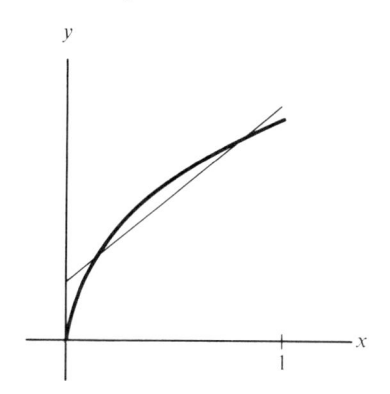

FIGURE 8-4

6. (c) Use part (b) with S, T replaced by $S^\perp$, $T^\perp$.
7. Parallel the proof of Theorem 8-3.
8. $f(t) = \pi + 2 \sin t$. (See Figure 8-5.)

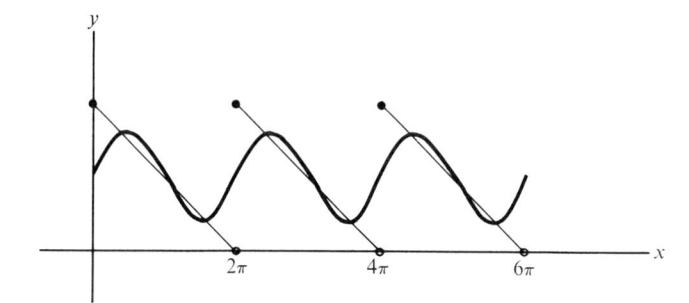

FIGURE 8-5

9. Let $\beta = \sigma + \tau$, where $\sigma = \sum_{i=1}^{n} (\beta, \alpha_i)\alpha_i$, and τ is orthogonal to σ (Theorems 8-2 and 8-3). Then Theorem 8-4 gives $\|\beta\|^2 = \|\sigma\|^2 + \|\tau\|^2$ and therefore $\|\beta\|^2 \geq \|\sigma\|^2 = \sum_{i=1}^{n} (\beta, \alpha_i)^2$. (See Example 13 of Lesson 12.)

10. $\beta = \sigma + \tau = (1, 0, 1) + (0, i, 0)$.

9

MATRICES AND MATRIX OPERATIONS

In Lesson 5 we used matrices in solving systems of linear equations. Now we will concentrate on the study of systems of matrices under various operations such as "addition" and "multiplication."

We begin with illustrative examples and then continue with a careful development of the rules of matrix algebra. There will be a flurry of new notations and definitions for you to remember and there will be elementary properties for you to understand. Practicing with the examples provided by the text and with examples of your own will be invaluable to you. This will enable you to read and understand later lessons without frequently referring to pages in order to recall an elementary fact just at the time when you need to concentrate without distraction on a new concept.

Example 1. Let

$$z_1 = y_1 + y_2 + y_3, \quad y_1 = 3x_1 - x_2,$$
$$z_2 = y_1 + 2y_2 - y_3, \quad y_2 = x_1 + 4x_2,$$
$$y_3 = - x_2.$$

The z's are given in terms of the y's, which are given in terms of the x's. We have compositely the z's in terms of the x's by

$$z_1 = 4x_1 + 2x_2$$
$$z_2 = 5x_1 + 8x_2$$

From the above we see three matrices

$$A = \begin{bmatrix} 1 & 1 & 1 \\ 1 & 2 & -1 \end{bmatrix} \quad B = \begin{bmatrix} 3 & -1 \\ 1 & 4 \\ 0 & -1 \end{bmatrix} \quad C = \begin{bmatrix} 4 & 2 \\ 5 & 8 \end{bmatrix}$$

Matrices and Matrix Operations

We show how to "multiply" A and B to get $AB = C$. The element in the *first* row and *first* column of AB is the *first* row of A dotted into the *first* column of B. The element in the *first* row and *second* column of AB is the *first* row of A dotted into the *second* column of B, etc. This "multiplication" gives $AB = C$.

Example 2. Let

$$A = \begin{bmatrix} 1 & -1 \\ 0 & 2 \\ 1 & 1 \end{bmatrix} \quad \text{and} \quad B = \begin{bmatrix} 2 & 1 & 2 & 0 \\ 3 & 2 & 1 & 1 \end{bmatrix}$$

Then

$$AB = \begin{bmatrix} -1 & -1 & 1 & -1 \\ 6 & 4 & 2 & 2 \\ 5 & 3 & 3 & 1 \end{bmatrix}$$

In Examples 1 and 2 we see that the product AB exists only if the number of columns of A equals the number of rows of B. Furthermore, if A is of type (m, n), that is, if A has m rows and n columns, and B is of type (n, p), then AB is of type (m, p).

Example 3. We also add matrices of the same type. Let

$$A = \begin{bmatrix} 1 & 2 & 3 \\ 4 & 1 & 0 \end{bmatrix} \quad B = \begin{bmatrix} 2 & 3 & 1 \\ 5 & -2 & 2 \end{bmatrix}$$

Then

$$A + B = \begin{bmatrix} 3 & 5 & 4 \\ 9 & -1 & 2 \end{bmatrix}$$

given by adding corresponding elements of A and B.

Finally, select a scalar, say 5. Then

$$5A = \begin{bmatrix} 5 & 10 & 15 \\ 20 & 5 & 0 \end{bmatrix}$$

the result of multiplying each element of A by 5.

Let us now explain formally what is meant by these matrix operations. We repeat the definition of a matrix, given earlier in Lesson 5.

Lessons in Linear Algebra

9-1 Definition

The rectangular array

$$A = \begin{bmatrix} a_{11} & a_{12} & \cdots & a_{1n} \\ a_{21} & a_{22} & \cdots & a_{2n} \\ & & \vdots & \\ a_{m1} & a_{m2} & \cdots & a_{mn} \end{bmatrix}$$

is called a *matrix of type* (m, n), and is abbreviated as $A = [a_{ij}]$ or as $A = [a_{ij}]_{m,n}$, where a_{ij} is the element in the ith row and jth column.

The matrix A is called *real (complex)* if all of its elements are real (complex).

Two matrices of the same type are *equal* if their corresponding elements are equal. That is, let $A = [a_{ij}]_{m,n}$, $B = [b_{ij}]_{m,n}$. Then $A = B$ means $a_{ij} = b_{ij}$ for $i = 1, 2, \ldots, m, j = 1, 2, \ldots, n$.

In this course we will concentrate on complex and real matrices, and you may assume that each matrix we consider is real or complex, unless there is a specific indication otherwise.

9-2 Definition

Let A and B be matrices, each of type (m, n). The *sum* $A + B$ is defined as the matrix of type (m, n) obtained by adding corresponding elements of A and B. That is, $A + B = [a_{ij} + b_{ij}]_{m,n}$.

If k is any scalar, we define the *scalar multiple* kA to be the matrix obtained by multiplying each element of A by k. That is,

$$kA = [ka_{ij}]_{m,n}$$

9-3 Definition

Let $A = [a_{ij}]_{m,n}$ and $B = [b_{ij}]_{n,p}$. Then the product AB is defined by

$$AB = [c_{ij}]_{m,p} \quad \text{where} \quad c_{ij} = \sum_{k=1}^{n} a_{ik}b_{kj},$$

the dot product of the ith row of A with the jth column of B.

The rules of matrix algebra are similar to those for real numbers, but there are important differences that must be remembered. For example, if A and B are "conformable" for addition, then $A + B = B + A$, and $k(A + B) = kA + kB$. Matrix multiplication is not in general commutative, even if A and B are "conformable" for multiplication. In fact AB can exist while BA does not.

Example 4. We illustrate the properties of matrix multiplication by taking various products of the matrices below.

$$A = \begin{bmatrix} 1 & 1 & 0 \\ 1 & 2 & 1 \end{bmatrix} \quad B = \begin{bmatrix} 1 \\ 3 \\ 2 \end{bmatrix} \quad C = [1, 0, 1, 0]$$

$$D = \begin{bmatrix} 2 & 1 \\ 1 & 2 \end{bmatrix} \quad E = \begin{bmatrix} 1 & 2 \\ 1 & 0 \end{bmatrix} \quad F = \begin{bmatrix} 3 & 2 \\ 1 & 2 \end{bmatrix}$$

Then

$$(AB)C = \begin{bmatrix} 4 \\ 9 \end{bmatrix} [1 \ 0 \ 1 \ 0] = \begin{bmatrix} 4 & 0 & 4 & 0 \\ 9 & 0 & 9 & 0 \end{bmatrix}$$

$$A(BC) = \begin{bmatrix} 1 & 1 & 0 \\ 1 & 2 & 1 \end{bmatrix} \begin{bmatrix} 1 & 0 & 1 & 0 \\ 3 & 0 & 3 & 0 \\ 2 & 0 & 2 & 0 \end{bmatrix} = \begin{bmatrix} 4 & 0 & 4 & 0 \\ 9 & 0 & 9 & 0 \end{bmatrix}$$

That the associative law $(AB)C = A(BC)$ holds generally will be proved shortly.

$$DE = \begin{bmatrix} 3 & 4 \\ 3 & 2 \end{bmatrix} \quad ED = \begin{bmatrix} 4 & 5 \\ 2 & 1 \end{bmatrix}$$

so $DE \neq ED$ even though both products exist.

$$EF = \begin{bmatrix} 5 & 6 \\ 3 & 2 \end{bmatrix} \quad FE = \begin{bmatrix} 5 & 6 \\ 3 & 2 \end{bmatrix}$$

Here the commutative law $EF = FE$ holds. Our conclusion is that commutativity of matrix multiplication does not hold generally but does hold in some cases.

As the last part of the example, consider the distributive laws $D(E + F) = DE + DF$ and $(E + F)D = ED + FD$.

$$D(E + F) = \begin{bmatrix} 2 & 1 \\ 1 & 2 \end{bmatrix} \begin{bmatrix} 4 & 4 \\ 2 & 2 \end{bmatrix} = \begin{bmatrix} 10 & 10 \\ 8 & 8 \end{bmatrix}$$

$$DE + DF = \begin{bmatrix} 3 & 4 \\ 3 & 2 \end{bmatrix} + \begin{bmatrix} 7 & 6 \\ 5 & 6 \end{bmatrix} = \begin{bmatrix} 10 & 10 \\ 8 & 8 \end{bmatrix}$$

Also

$$(E + F)D = \begin{bmatrix} 4 & 4 \\ 2 & 2 \end{bmatrix} \begin{bmatrix} 2 & 1 \\ 1 & 2 \end{bmatrix} = \begin{bmatrix} 12 & 12 \\ 6 & 6 \end{bmatrix}$$

$$ED + FD = \begin{bmatrix} 4 & 5 \\ 2 & 1 \end{bmatrix} + \begin{bmatrix} 8 & 7 \\ 4 & 5 \end{bmatrix} = \begin{bmatrix} 12 & 12 \\ 6 & 6 \end{bmatrix}$$

It is generally true that both distributive laws hold, but $D(E + F)$ may not equal $(E + F)D$.

Lessons in Linear Algebra

9-4 Theorem

Let A, B, C represent general matrices and let a, b be general scalars. The following laws hold, provided the indicated sums and products exist:

(1) $(A + B) + C = A + (B + C)$.
(2) $A + B = B + A$.
(3) $(AB)C = A(BC)$.
(4) $A(B + C) = AB + AC$ and $(B + C)A = BA + CA$.
(5) $(a + b)A = aA + bA$.
(6) $a(A + B) = aA + aB$.
(7) $(aA)(bB) = abAB$.

Proof: We prove (1), (3), and (6) and leave the others for the reader.
(1) Let $A = [a_{ij}]$, $B = [b_{ij}]$, and $C = [c_{ij}]$ be of type (m, n). Then

$A + B = [a_{ij} + b_{ij}]$ (definition of addition)
$(A + B) + C = [a_{ij} + b_{ij} + c_{ij}]$ (definition of addition)
$= A + (B + C)$ (definition of addition)

The student may feel that here we have proved the obvious. Even so, you should notice the notation. In the first step the statement $A + B = [a_{ij} + b_{ij}]$ means that the ij-element of $A + B$ is $a_{ij} + b_{ij}$, the sum of the corresponding elements of A and B.

(3) The associative law $(AB)C = A(BC)$ will be proved directly from the definition of matrix multiplication, but a more elegant proof will be outlined later, a proof which identifies each matrix with a mapping and then uses the fact that composition of mappings is always associative.

Let $A = [a_{ij}]_{m,n}$, $B = [b_{ij}]_{n,p}$, $C = [c_{ij}]_{p,r}$. Using only the definition of matrix multiplication and the properties of summations, we have

$$AB = \left[\sum_{k=1}^{n} a_{ik} b_{kj} \right]_{m,p}$$

$$(AB)C = \left[\sum_{h=1}^{p} \left(\sum_{k=1}^{n} a_{ik} b_{kh} \right) c_{hj} \right]_{m,r}$$

On the other hand

$$(BC) = \left[\sum_{h=1}^{p} b_{ih} c_{hj} \right]_{n,r}$$

$$A(BC) = \left[\sum_{k=1}^{n} a_{ik} \left(\sum_{h=1}^{p} b_{kh} c_{hj} \right) \right]_{m,r}$$

$$= \left[\sum_{k=1}^{n} \sum_{h=1}^{p} a_{ik} b_{kh} c_{hj} \right]_{m,r} = (AB)C$$

Matrices and Matrix Operations

(6) Let A and B be of type (m, n). Then

$$\begin{aligned} a(A + B) &= a[a_{ij} + b_{ij}] \quad \text{(definition of addition)} \\ &= [aa_{ij} + ab_{ij}] \quad \text{(definition of scalar multiplication)} \\ &= [aa_{ij}] + [ab_{ij}] \quad \text{(definition of addition)} \\ &= aA + aB \quad \text{(definition of scalar multiplication)} \quad \blacksquare \end{aligned}$$

We introduce now some very special matrices. The matrix A is called a "zero matrix" if each element of A is zero. Such a matrix will be denoted 0, but it is understood that there are many such zero matrices, one for each type (m, n). It is obvious that $A + 0 = 0 + A = A$ and that $A0 = 0A = 0$ whenever the indicated sums and products exist. Of course, a matrix equation such as $A0 = 0$ involves two zero matrices which may be of different types. For example

$$\begin{bmatrix} 1 & 2 & 3 \\ 4 & 1 & 2 \end{bmatrix} \begin{bmatrix} 0 & 0 \\ 0 & 0 \\ 0 & 0 \end{bmatrix} = \begin{bmatrix} 0 & 0 \\ 0 & 0 \end{bmatrix}$$

If $A + B = 0$, then B is called the *negative* of A, denoted $-A$. Clearly $-A = (-1)A$. Subtraction is defined by $A - B = A + (-B)$.

Another special matrix of importance is the "identity" matrix I, a square matrix with each diagonal element equal to 1 and each off-diagonal element equal to 0. The identity matrix of order 2 is

$$\begin{bmatrix} 1 & 0 \\ 0 & 1 \end{bmatrix},$$

the identity matrix of order 3 is

$$\begin{bmatrix} 1 & 0 & 0 \\ 0 & 1 & 0 \\ 0 & 0 & 1 \end{bmatrix}, \text{etc.}$$

Whenever it is important to designate the order n of the identity matrix I we shall use I_n.

The reader should verify that $AI = A$ and $IA = A$, if the products exist.

Let A, B, C be any matrices and let a be a general scalar. Then

(1) $-(aA) = (-a)A$.
(2) $AI = A$, $IA = A$ if the products exist.
(3) $A + B = A + C$ implies $B = C$ (a cancellation law).

9-5
Theorem

Lessons in Linear Algebra

Proof: (1) and (2) are left as exercises. For (3),

$$A + B = A + C \Rightarrow -A + (A + B) = -A + (A + C)$$
(existence of $-A$, uniqueness of addition)
$$\Rightarrow (-A + A) + B = (-A + A) + C$$
(associative law)
$$\Rightarrow 0 + B = 0 + C$$
$(-A + A = 0)$
$$\Rightarrow B = C$$
(property of zero matrix) ∎

The exercises which follow are designed to help you understand the properties of matrix operations and to give you practice with these operations. As you consider the exercises, try to get by with as little writing as you can. For example, do not write out an m by n matrix A if you can simply refer to it as A or as $[a_{ij}]$. Use the rules of matrix algebra in their simple form whenever you can.

Summary of Topics in Lesson 9

Matrix operations. Multiplication, addition, scalar multiplication.
Notation $[a_{ij}]_{m,n}$ for a matrix of type m, n.
Rules of matrix algebra.
Special matrices. The zero matrix. The identity matrix.
The cancellation law $A + B = A + C \Rightarrow B = C$.

Exercises

1. Given

$$A = \begin{bmatrix} 1 & 2 \\ -1 & 0 \\ 4 & 3 \end{bmatrix} \quad B = \begin{bmatrix} 1 & 0 & 1 & 0 \\ 0 & 1 & 1 & 1 \end{bmatrix} \quad C = \begin{bmatrix} 1 & 0 & 0 \\ 0 & 1 & 0 \\ 0 & 0 & 1 \\ 1 & 1 & 1 \end{bmatrix}$$

verify that $(AB)C = A(BC)$ by performing all of the indicated matrix multiplications.

2. (a) Let

$$A = \begin{bmatrix} 1 & 4 \\ 1 & 3 \end{bmatrix} \quad \text{and} \quad B = \begin{bmatrix} -3 & 4 \\ 1 & -1 \end{bmatrix}$$

and verify that $AB = BA = I$. (Whenever this occurs A and B are said to be "inverse" matrices and we write $B = A^{-1}$ or $A = B^{-1}$.)

(b) Let
$$A = \begin{bmatrix} 1 & 2 & 1 \\ 1 & 4 & 2 \end{bmatrix} \quad B = \begin{bmatrix} 2 & -1 \\ -1 & 2 \\ 1 & -3 \end{bmatrix}$$

Verify that AB is an identity matrix but BA is not. (A is a "left-inverse" of B, and B is a "right-inverse" of A. We do not say that A, B are inverse matrices.)

3. In the algebra of real numbers $xy = 0$ implies either x or $y = 0$. Investigate matrix multiplication and determine whether or not $AB = 0$ implies either $A = 0$ or $B = 0$. Either prove the implication holds for all matrices A, B (for which AB is defined), *or else* display a counterexample, that is, find a nonzero pair of matrices whose product is zero.

4. Prove a generalized associative law for the product of n matrices $A_1, \ldots, A_n$ by showing that any such product can be expressed $A_1(A_2(A_3 \ldots (A_{n-1}A_n) \ldots)$.

5. By A^2 we mean AA, by A^3 we mean AAA. Formally, if A is square, A^0 is defined as I, and for n a positive integer, A^n is defined as $A^{n-1}A$.
   (a) Given
   $$A = \begin{bmatrix} 2 & 3 \\ 1 & 1 \end{bmatrix} \quad \text{and} \quad B = \begin{bmatrix} 1 & 0 \\ 1 & 1 \end{bmatrix}$$
   compute A^2 and B^2. Is it true that $A^2 - B^2 = (A + B)(A - B)$?
   (b) Under what circumstances do two matrices A and B satisfy the matrix equation $A^2 - B^2 = (A + B)(A - B)$?

6. Given
$$A = \begin{bmatrix} 1 & x \\ 0 & 1 \end{bmatrix}$$
compute A^n.

7. Let A be an n by n matrix, each element of which is 1. Compute A^m, where m is a positive integer.

8. Let $A = [a_{ij}]$ be an nth order square matrix such that $a_{ij} = (-1)^{i+j}$ for all i, j. Find A^2 and then show by matrix algebra that $A(A - nI) = 0$. Also, show that for any positive integer m, $A^m = n^{m-1}A$.

Lessons in Linear Algebra

9. Let A be an nth order square matrix with typical element $a_{ij} = i$. Express A^m as a multiple of A, given that m is a positive integer.

10. (a) Let
$$A = \begin{bmatrix} 1 & 2 \\ 2 & 4 \end{bmatrix} \quad B = \begin{bmatrix} 3 & 0 \\ 1 & 1 \end{bmatrix} \quad C = \begin{bmatrix} 1 & 2 \\ 2 & 0 \end{bmatrix}$$
Verify that $AB = AC$. What does this assert about the cancellation law $AB = AC \Rightarrow B = C$?

(b) Show that if A has a left inverse D such that $DA = I$, then the cancellation law $AB = AC \Rightarrow B = C$ holds.

11. For
$$A = \begin{bmatrix} a & b \\ c & d \end{bmatrix}$$
compute $A^2 - (a+d)A + (ad - bc)I$.

12. Five cities A, B, C, D, E are connected by airlines according to the diagram in Figure 9-1. $B \rightarrow C$ means there is a flight from B to C,

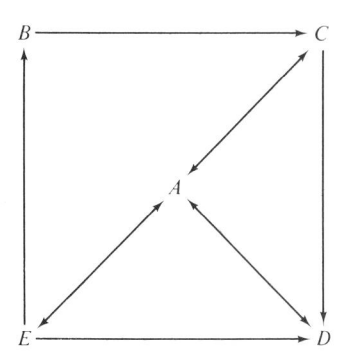

FIGURE 9-1

$A \leftrightarrow C$ means there are flights from A to C and from C to A, etc. To represent the situation we construct an "incidence" matrix of type (5, 5).

$$\begin{array}{c} \\ A \\ B \\ C \\ D \\ E \end{array} \begin{array}{c} \begin{array}{ccccc} A & B & C & D & E \end{array} \\ \begin{bmatrix} 0 & 0 & 1 & 1 & 1 \\ 0 & 0 & 1 & 0 & 0 \\ 1 & 0 & 0 & 1 & 0 \\ 1 & 0 & 0 & 0 & 0 \\ 1 & 1 & 0 & 1 & 0 \end{bmatrix} \end{array}$$

Matrices and Matrix Operations

Interpret this matrix. Construct an incidence matrix M to represent the diagram in Figure 9-2. Let M^T denote "M-transpose," the

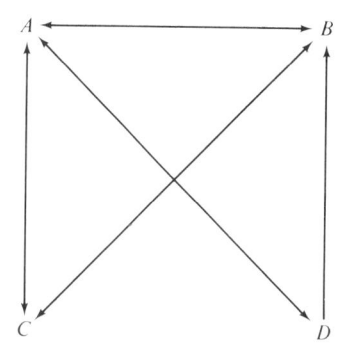

FIGURE 9-2

matrix obtained by interchanging the rows and columns of M. What numbers appear along the diagonal of the matrix MM^T?

13. A chain of four stores report receipts six days per week. The report from store number 1 would be summarized at the end of the week with $[r_{11} \; r_{12} \; r_{13} \; r_{14} \; r_{15} \; r_{16}]$, where r_{13} represents the receipts at store number 1 on the third day, Wednesday, etc. The other stores submit the same report, so the entire receipt picture is given by the 4 by 6 matrix $R = [r_{ij}]$. A 4 by 6 "cost" matrix $C = [c_{ij}]$ is also constructed, where c_{ij} is the cost of operating store number i on the jth day. Let $P = R - C$, and interpret $P = [p_{ij}]$. Then let $a_{ij} = 100(p_{ij}/c_{ij})$ and interpret the matrix A. Let U be the matrix of type $(6,1)$ with each element equal to 1, and let V be the matrix of type $(1, 4)$, each of whose elements is 1. Interpret the nine matrices RU, CU, PU, VR, VC, VP, VRU, VCU, VPU.

14. The attitudes of five varsity players A, B, C, D, E toward each other are tested by asking each one to describe his feeling toward each of the others. The numbers $1, -1, 0$ are recorded to show whether the feeling is favorable, unfavorable, or neutral. A matrix M is constructed.

$$\begin{array}{c} \\ A \\ B \\ C \\ D \\ E \end{array} \begin{array}{c} \begin{array}{ccccc} A & B & C & D & E \end{array} \\ \begin{bmatrix} 0 & 1 & 1 & 1 & 1 \\ 0 & 0 & -1 & 1 & -1 \\ 1 & 0 & 0 & -1 & -1 \\ 1 & 1 & -1 & 0 & 0 \\ 1 & 1 & -1 & 0 & 0 \end{bmatrix} \end{array}$$

Lessons in Linear Algebra

The entry $m_{12} = 1$ indicates that A likes B. The entry $m_{21} = 0$ means that B is neutral toward A. The entry $m_{43} = -1$ means that D dislikes C. Let U and V be matrices with each element 1, U of type (5, 1), V of type (1, 5). Compute and interpret MU, VM, and VMU.

Remarks on Exercises

1. $(AB)C = A(BC) = \begin{bmatrix} 3 & 4 & 5 \\ -1 & 0 & -1 \\ 7 & 6 & 10 \end{bmatrix}$.

4. Use mathematical induction. The law holds for $n = 3$, since $(A_1 A_2)A_3 = A_1(A_2 A_3)$. Assume the law holds for any product of k matrices, where $k < n$. This means we can write the product $A_1 A_2 \ldots A_k$ without indicating the order in which the products are taken. Now any product of the n matrices is the product of two matrices

$$P = [A_1 \ldots A_k][A_{k+1} \ldots A_n]$$

where the induction hypothesis permits the free association of the two products enclosed in brackets. The associative law for three matrices gives

$$P = A_1 [A_2 \ldots A_k][A_{k+1} \ldots A_n]$$

and the induction hypothesis gives

$$P = A_1(A_2(A_3 \ldots \ldots (A_{n-1} A_n) \ldots)$$

5. (a) $A^2 - B^2 = \begin{bmatrix} 6 & 9 \\ 1 & 3 \end{bmatrix}$. $(A + B)(A - B) = \begin{bmatrix} 3 & 9 \\ 2 & 6 \end{bmatrix}$.

   (b) When $AB = BA$.

6. $\begin{bmatrix} 1 & nx \\ 0 & 1 \end{bmatrix}$.

7. $n^{m-1}A$.

8. First verify that $A^2 = nA$. Factoring $A^2 - nA = 0$ gives $A(A - nI) = 0$. Use mathematical induction to show that $A^m = n^{m-1}A$.

9. The sum $1 + 2 + \ldots + n = (n/2)(n + 1)$. Then

$$A^2 = \left(\frac{n}{2}\right)(n + 1)A$$

Thus,
$$A^m = \left[\left(\frac{n}{2}\right)(n+1)\right]^{m-1} A$$

10. (a) The cancellation law does not hold.
    (b) Multiply both sides of $AB = AC$ by D.

11. 0.

12. $M = \begin{bmatrix} 0 & 1 & 1 & 1 \\ 1 & 0 & 1 & 0 \\ 1 & 1 & 0 & 0 \\ 1 & 1 & 0 & 0 \end{bmatrix}$.

    The numbers 3, 2, 2, 2 along the diagonal of MM^T represent the number of "flights", respectively, from A, B, C, D.

13. P is a "profit" matrix, where p_{ij} is the profit from the ith store on the jth day. The matrix A represents a percentage profit (relative to the cost). The four elements of RU give the weekly receipts for each of the four stores. The six elements of VR give the total of the six daily receipts turned in from all the stores. The single element of VRU gives the grand total of the weekly receipts.

14. $MU = \begin{bmatrix} 4 \\ -1 \\ -1 \\ 1 \\ 1 \end{bmatrix}$. $VM = \begin{bmatrix} 3 & 3 & -2 & 1 & -1 \end{bmatrix}$. $VMU = [4]$.

10

SOME USES OF MATRIX NOTATION AND MATRIX ALGEBRA

Much of the remainder of the entire text uses matrix algebra. This lesson is a brief introduction to several ways in which matrix notation and matrix algebra are used in solving problems. The details of just why a notation or process applies are left for a later lesson; right now you may observe certain matrices being pulled out of hats, and if you are curious enough you will ponder just how the choice is made, and you will look forward to a general treatment which explains and validates the choice.

Example 1. The use of matrix algebra in systems of linear equations. Consider the system of equations

(10-1)
$$3x_1 + 13x_2 + 5x_3 = 1$$
$$x_1 + 4x_2 + 2x_3 = -2$$

and let $X = [x_1 \ x_2 \ x_3]$ be the 1 by 3 matrix of the unknowns.* The system can be written in several ways:

$$\begin{bmatrix} 3 & 13 & 5 \\ 1 & 4 & 2 \end{bmatrix} \begin{bmatrix} x_1 \\ x_2 \\ x_3 \end{bmatrix} = \begin{bmatrix} 1 \\ -2 \end{bmatrix}$$

or

(10-2)
$$AX^T = B$$

where

$$A = \begin{bmatrix} 3 & 13 & 5 \\ 1 & 4 & 2 \end{bmatrix} \quad X^T = \begin{bmatrix} x_1 \\ x_2 \\ x_3 \end{bmatrix} \quad B = \begin{bmatrix} 1 \\ -2 \end{bmatrix}$$

---

*We will use the notations $[x_1 x_2 \ldots x_n]$ and $(x_1, x_2, \ldots, x_n)$ interchangeably to represent a 1 by n matrix or n-tuple.

Some Uses of Matrix Notation and Matrix Algebra

The matrix equation $AX^T = B$ is an excellent way to write the system as a matrix equation with the unknown X.

The system can also be written

$$x_1 \begin{bmatrix} 3 \\ 1 \end{bmatrix} + x_2 \begin{bmatrix} 13 \\ 4 \end{bmatrix} + x_3 \begin{bmatrix} 5 \\ 2 \end{bmatrix} = \begin{bmatrix} 1 \\ -2 \end{bmatrix}$$

or

$$x_1 C_1 + x_2 C_2 + x_3 C_3 = B \quad (10\text{-}3)$$

where C_1 is the first column of A, etc. Here we see the problem as one of expressing the vector B in R^2 as a linear combination of C_1, C_2, C_3.

Another way of writing the system is

$$(3, 13, 5) \cdot (x_1, x_2, x_3) = 1$$
$$(1, 4, 2) \cdot (x_1, x_2, x_3) = -2$$

or

$$R_1 \cdot X = 1$$
$$R_2 \cdot X = -2 \quad (10\text{-}4)$$

where R_1 is the first row of A, etc. Here we see the problem as one of finding a vector X so that its dot products with R_1 and R_2 are 1 and -2, respectively.

It is rarely desirable to represent the system in so many different ways, but there are examples which make good use of each representation.

Now premultiply (from the left) each side of Equation (10-2) by the *carefully selected* matrix

$$D = \begin{bmatrix} -4 & 13 \\ 1 & -3 \end{bmatrix}$$

This gives $DAX = DB$ or

$$\begin{bmatrix} 1 & 0 & 6 \\ 0 & 1 & -1 \end{bmatrix} \begin{bmatrix} x_1 \\ x_2 \\ x_3 \end{bmatrix} = \begin{bmatrix} -30 \\ 7 \end{bmatrix}$$

the original system in Hermite normal form. Hence

$$x_1 + 6x_3 = -30$$
$$x_2 - x_3 = 7$$

and the general solution is the set

$$\{x_3(-6, 1, 1) + (-30, 7, 0) \mid x_3 \in R\}$$

Lessons in Linear Algebra

Pick a solution at random, for example $X = (-30, 7, 0)$. For this solution, Equation (10-3) is checked by

$$-30C_1 + 7C_2 = -30\begin{bmatrix}3\\1\end{bmatrix} + 7\begin{bmatrix}13\\4\end{bmatrix} = \begin{bmatrix}1\\-2\end{bmatrix}$$

and (10-4) is checked by

$$(3, 13, 5) \cdot (-30, 7, 0) = 1$$
$$(1, 4, 2) \cdot (-30, 7, 0) = -2$$

In general the system

$$\sum_{j=1}^{n} a_{ij}x_j = b_i, \quad i = 1, 2, \ldots, m$$

of m linear equations in n unknowns can be written

$$AX^T = B$$

or

$$x_1 C_1 + \ldots + x_n C_n = B$$

or

$$R_i \cdot X = b_i, \quad i = 1, 2, \ldots, m$$

We pause to make a brief definition, followed by a theorem.

10-5 Definition

Let $A = [a_{ij}]$ be a matrix of type (m, n). The *transpose of A*, denoted A^T, is the matrix obtained by interchanging rows and columns of A, respectively. That is $A^T = [b_{ij}]_{n,m}$, where $b_{ij} = a_{ji}$. (It is obvious that $(A^T)^T = A$.)

To illustrate, let

$$A = \begin{bmatrix}1 & 2\\3 & 4\end{bmatrix} \quad \text{and} \quad B = \begin{bmatrix}1 & 1 & 0\\0 & 2 & 1\end{bmatrix}$$

Then

$$A^T = \begin{bmatrix}1 & 3\\2 & 4\end{bmatrix} \quad \text{and} \quad B^T = \begin{bmatrix}1 & 0\\1 & 2\\0 & 1\end{bmatrix}$$

The matrix AB exists, and, therefore, $(AB)^T$ exists. However, $A^T B^T$ does not exist so it would be folly to claim that $(AB)^T = A^T B^T$.

It does turn out that $(AB)^T = B^T A^T$, a fact that we now generalize.

Some Uses of Matrix Notation and Matrix Algebra

If the matrix product AB exists, then $(AB)^T = B^T A^T$.

10-6 Theorem

Proof: Let $A = [a_{ij}]_{m,n}$ and $B = [b_{ij}]_{n,p}$. Then $(AB)^T$ is of type (p, m) and so is $B^T A^T$. The ij-element of $(AB)^T$ is the ji-element of AB, specifically $\sum_{k=1}^{n} a_{jk} b_{ki}$. On the other hand, the ij-element of $B^T A^T = \sum_{k=1}^{n}$ (ik-element of B^T)(kj-element of A^T) $= \sum_{k=1}^{n} b_{ki} a_{jk}$. Therefore, $(AB)^T = B^T A^T$. ∎

Example 2. It is easy to represent the dot product in R^n as the product of matrices. Let $X = [x_1, x_2, \ldots, x_n]$ and $Y = [y_1, y_2, \ldots, y_n]$ be two vectors in R^n. Then

$$X \cdot Y = XY^T$$

You should practice interpreting the rules governing the dot product as rules in matrix algebra. As an illustration, we check that $X \cdot Y = Y \cdot X$. Since XY^T is a 1 by 1 matrix, it is its own transpose. Consequently, $X \cdot Y = XY^T = (XY^T)^T = (Y^T)^T X^T = YX^T = Y \cdot X$.

Let A be a square matrix. A is *symmetric* if $A^T = A$. A is *skew-symmetric* if $A^T = -A$.

10-7 Definition

The matrix

$$\begin{bmatrix} 1 & 7 & 2 \\ 7 & -4 & 5 \\ 2 & 5 & 3 \end{bmatrix}$$

is symmetric. The matrix

$$\begin{bmatrix} 0 & 7 & -3 \\ -7 & 0 & 4 \\ 3 & -4 & 0 \end{bmatrix}$$

is skew-symmetric. For A to be symmetric we must have each $a_{ij} = a_{ji}$. The condition for skew-symmetry is $a_{ij} = -a_{ji}$, requiring among other things that each diagonal element $a_{ii} = 0$.

It is easy to show that the sum of two symmetric matrices (of the same order) is symmetric, but the product of two symmetric matrices is not always symmetric. Similar statements hold for skew-symmetric matrices.

Lessons in Linear Algebra

10-8 Theorem

Any square matrix A can be expressed as the sum of two matrices B and C, where B is symmetric and C is skew-symmetric. Furthermore, the matrices B and C are unique.

Proof: We set out to find matrices B, C such that $B^T = B$, $C^T = -C$ and $A = B + C$. Transposing each member of the last equation gives $A^T = B - C$. Therefore, $A + A^T = 2B$, $A - A^T = 2C$, and the desired matrices are $B = (1/2)(A + A^T)$, $C = (1/2)(A - A^T)$. The matrix B is symmetric, C is skew-symmetric. The steps above show also that B and C are unique.

To illustrate the theorem,

$$\begin{bmatrix} 1 & 2 & 3 \\ -4 & 2 & -5 \\ 1 & -1 & 7 \end{bmatrix} = \begin{bmatrix} 1 & -1 & 2 \\ -1 & 2 & -3 \\ 2 & -3 & 7 \end{bmatrix} + \begin{bmatrix} 0 & 3 & 1 \\ -3 & 0 & -2 \\ -1 & 2 & 0 \end{bmatrix}$$

that is,

$$A = \frac{1}{2}(A + A^T) + \frac{1}{2}(A - A^T) \quad \blacksquare$$

Example 3. The use of matrices in representing bilinear and quadratic forms. We start here with a special example of a "bilinear" form,

$$x_1 y_1 + 2x_1 y_2 + 3x_1 y_3$$
$$- 4x_2 y_1 + 2x_2 y_2 - 5x_2 y_3$$
$$+ x_3 y_1 - x_3 y_2 + 7x_3 y_3$$

which can be written as

$$\begin{bmatrix} x_1 & x_2 & x_3 \end{bmatrix} \begin{bmatrix} 1 & 2 & 3 \\ -4 & 2 & -5 \\ 1 & -1 & 7 \end{bmatrix} \begin{bmatrix} y_1 \\ y_2 \\ y_3 \end{bmatrix}$$

or as XAY^T. The form is linear in $X = \begin{bmatrix} x_1 & x_2 & x_3 \end{bmatrix}$ and linear in $Y = \begin{bmatrix} y_1 & y_2 & y_3 \end{bmatrix}$.

Whenever X and Y are equal, we call the form a "quadratic" form. The quadratic form XAX^T can be written out as

$$\begin{bmatrix} x_1 & x_2 & x_3 \end{bmatrix} \begin{bmatrix} 1 & 2 & 3 \\ -4 & 2 & -5 \\ 1 & -1 & 7 \end{bmatrix} \begin{bmatrix} x_1 \\ x_2 \\ x_3 \end{bmatrix}$$

or as $x_1^2 + 2x_2^2 + 7x_3^2 - 2x_1 x_2 + 4x_1 x_3 - 6x_2 x_3$.

However, the same quadratic form XAX^T can be represented just as precisely with many other matrices B so that XBX^T is identically the same as XAX^T. The same form is given by

$$[x_1 \quad x_2 \quad x_3] \begin{bmatrix} 1 & 3 & 4 \\ -5 & 2 & -4 \\ 0 & -2 & 7 \end{bmatrix} \begin{bmatrix} x_1 \\ x_2 \\ x_3 \end{bmatrix}$$

or by

$$[x_1 \quad x_2 \quad x_3] \begin{bmatrix} 1 & -1 & 2 \\ -1 & 2 & -3 \\ 2 & -3 & 7 \end{bmatrix} \begin{bmatrix} x_1 \\ x_2 \\ x_3 \end{bmatrix}$$

where the last matrix is symmetric, equal to $(1/2)(A + A^T)$. The representation using the symmetric matrix is more useful than the others.

The quadratic form XAX^T in the n variables $X = [x_1 \quad x_2 \quad \ldots \quad x_n]$ can be written uniquely as XBX^T where B is symmetric.

10-9 Theorem

Proof: Let $A = B + C$, where B is symmetric and C skew-symmetric. The 1 by 1 matrix XCX^T is symmetric, and since C is skew-symmetric, we see that $XCX^T = (XCX^T)^T = X^{TT}C^TX^T = -XCX^T$, and $XCX^T = 0$ for all X.

We now turn to the given quadratic form.

$$XAX^T = X(B + C)X^T = XBX^T + XCX^T = XBX^T$$

where B is symmetric.

The proof that the quadratic form XAX^T has just one representation XBX^T with B symmetric is left for you. The proof is outlined in Exercise 11. ∎

Example 4. Representing systems of differential equations with matrices. Consider the following system of differential equations.

$$\frac{dy_1}{dx} = a_{11}y_1 + a_{12}y_2 + \ldots + a_{1n}y_n + b_1$$

$$\frac{dy_2}{dx} = a_{21}y_1 + a_{22}y_2 + \ldots + a_{2n}y_n + b_2$$

$$\vdots$$

$$\frac{dy_n}{dx} = a_{n1}y_1 + a_{n2}y_2 + \ldots + a_{nn}y_n + b_n$$

Lessons in Linear Algebra

The problem is to find $y_1, y_2, \ldots, y_n$, each a function of x so that the system is satisfied, given that each b_i is a continuous function of x. Let $Y = [y_1 \ \ldots \ y_n]$, $A = [a_{ij}]_{n,n}$, $B = [b_1 \ \ldots \ b_n]$. Then the system can be written

$$\frac{d}{dx}\begin{bmatrix} y_1 \\ \vdots \\ y_n \end{bmatrix} = \begin{bmatrix} a_{11} & \cdots & a_{1n} \\ \vdots & & \vdots \\ a_{n1} & \cdots & a_{nn} \end{bmatrix} \begin{bmatrix} y_1 \\ \vdots \\ y_n \end{bmatrix} + \begin{bmatrix} b_1 \\ \vdots \\ b_n \end{bmatrix}$$

or as

$$\frac{dY^T}{dx} = AY^T + B^T$$

We illustrate further with a two by two system,

$$y_1' = 2y_1 + 3y_2$$
$$y_2' = 4y_1 + 3y_2$$

represented in matrix form by

(10-10)
$$\frac{dY^T}{dx} = AY^T$$

where

$$A = \begin{bmatrix} 2 & 3 \\ 4 & 3 \end{bmatrix} \quad \text{and} \quad Y = [y_1 \ y_2]$$

Let

$$P = \begin{bmatrix} 1 & 3 \\ -1 & 4 \end{bmatrix}$$

and change the dependent variables $Y = [y_1 \ y_2]$ to $Z = [z_1 \ z_2]$ by means of

$$Y^T = PZ^T$$

The system (10-10) becomes

$$P\frac{dZ^T}{dx} = APZ^T$$

Now multiply both sides by

$$Q = \left(\frac{1}{7}\right)\begin{bmatrix} 4 & -3 \\ 1 & 1 \end{bmatrix}$$

Some Uses of Matrix Notation and Matrix Algebra

a matrix for which $QP = I$. We have

$$\frac{dZ^T}{dx} = QAPZ^T$$

or

$$\frac{d}{dx}\begin{bmatrix} z_1 \\ z_2 \end{bmatrix} = \begin{bmatrix} -1 & 0 \\ 0 & 6 \end{bmatrix}\begin{bmatrix} z_1 \\ z_2 \end{bmatrix}$$

or

$$z_1' = -z_1$$
$$z_2' = 6z_2$$

Knowing that $z_1 = c_1 e^{-x}$ and $z_2 = c_2 e^{6x}$ leads us through $Y^T = PZ^T$ to the solution

$$y_1 = c_1 e^{-x} + 3c_2 e^{6x}$$
$$y_2 = -c_1 e^{-x} + 4c_2 e^{6x}$$

Optional Section on Complex Matrices

Complex matrices are used in systems of linear equations in the same way that real matrices are used. The proofs of the rules of matrix algebra such as $(AB)C = A(BC)$, $(AB)^T = B^T A^T$, etc., apply to complex matrices.

The concepts of symmetry and skew-symmetry are extended into the system of complex matrices by new definitions.

10-11 Definition

The square complex matrix A is *Hermitian* if $\bar{A}^T = A$. The matrix A is skew-Hermitian if $\bar{A}^T = -A$.

Let

$$A = \begin{bmatrix} 1 & i & 1-2i \\ -i & 3 & 2+i \\ 1+2i & 2-i & 5 \end{bmatrix} \quad B = \begin{bmatrix} 0 & i & 1+3i \\ i & 0 & 5+i \\ -1+3i & -5+i & 0 \end{bmatrix}$$

A is Hermitian, B skew-Hermitian.

A real Hermitian matrix is symmetric, and a real skew-Hermitian matrix is skew-symmetric.

It is customary to write $\bar{A}^T$ as A^*. Rules such as $(A + B)^* = A^* + B^*$, $(AB)^* = B^* A^*$, $A^{**} = A$ are useful.

We have the analog of Theorem 10-8.

111

Lessons in Linear Algebra

10-12 Theorem

Any square matrix A can be expressed $A = B + C$, where B is Hermitian and C skew-Hermitian. Also B and C are unique.

The proof parallels that of Theorem 10-8, with $B = (1/2)(A + A^*)$, $C = (1/2)(A - A^*)$.

Example 5. Let

$$A = \begin{bmatrix} 2+i & 4-i \\ 5i & 3 \end{bmatrix}$$

Then $A = B + C$, where

$$B = \frac{1}{2}(A + A^*) = \begin{bmatrix} 2 & 2-3i \\ 2+3i & 3 \end{bmatrix}$$

$$C = \frac{1}{2}(A - A^*) = \begin{bmatrix} i & 2+2i \\ -2+2i & 0 \end{bmatrix}$$

where B is Hermitian and C is skew-Hermitian.

The "complex" quadratic form XAX^* is

$$(2+i)x_1\bar{x}_1 + (4-i)x_1\bar{x}_2 + 5ix_2\bar{x}_1 + 3x_2\bar{x}_2$$

Summary of Topics in Lesson 10

The matrix notation in systems of linear equations.

Transpose of a matrix.

Symmetric matrices. Skew-symmetric matrices.

Bilinear and quadratic forms.

Systems of differential equations.

Optional topics on complex matrices: Hermitian and skew-Hermitian matrices.

Exercises

1. Given the system

$$x_1 - 2x_2 + x_3 - x_4 = 1$$
$$x_1 - x_2 + x_3 + 2x_4 = 2$$
$$x_1 - 4x_2 + x_3 - 7x_4 = 1$$

(a) Display matrices A, X, B so that the system is represented by the matrix equation $AX^T = B$.
(b) Premultiply each side $AX^T = B$ by the matrix
$$D = \begin{bmatrix} -1 & 2 & 0 \\ -1 & 1 & 0 \\ -3 & 2 & 1 \end{bmatrix}$$
and solve the system.

2. Let $X = [x_1 \quad x_2 \quad x_3]$ and
$$A = \begin{bmatrix} 1 & 1 & 2 & 1 \\ 1 & 2 & -1 & -3 \\ 2 & 3 & 1 & -2 \end{bmatrix}$$
Write out the system given by the matrix equation $XA = 0$, and solve the system.

3. Let A be a real n by n matrix with columns $C_1, C_2, \ldots, C_n$, $X = [x_1 \ldots x_n]$, and $B = [b_1 \ldots b_n]$.
  (a) What can you conclude about the solution of $AX^T = B^T$ if the columns of A form a basis of R^n? (Use the notation 10-3.)
  (b) What can you conclude about the solution of $AX^T = 0$ if the columns of A are linearly dependent vectors in R^n?

4. (a) Explain why any vector $X = (x_1, \ldots, x_n)$ which satisfies the matrix equation $AX^T = 0$ must be orthogonal to each row of A.
  (b) Interpret part (a) for the planes
$$2x + y - z = 0, \quad x + y - 5z = 0$$

5. Let
$$A = \begin{bmatrix} 1 & 3 & 7 \\ 2 & -1 & 4 \\ -3 & -6 & 5 \end{bmatrix}$$
  (a) Write out the quadratic form XAX^T, and evaluate it at $X = [1, -1, 1]$.
  (b) Express A as the sum of two matrices B and C, so that B is symmetric and C is skew-symmetric, and verify that XBX^T is the same quadratic form as XAX^T.

6. Find a symmetric matrix A such that XAX^T represents the quadratic form $x_1^2 + x_2^2 - x_4^2 + 2x_1x_2 - 6x_1x_3 + 8x_1x_4 + 4x_2x_3 - x_3x_4$.

7. (a) Given
$$A = \begin{bmatrix} 1 & 0 \\ 0 & 2 \end{bmatrix}$$
show that the quadratic form $XAX^T \geq 0$ and that $XAX^T = 0$ if and only if $X = 0$.
   (b) What can you say about the range of values of XBX^T, if
$$B = \begin{bmatrix} 1 & 0 & 0 \\ 0 & 2 & 0 \\ 0 & 0 & 0 \end{bmatrix}?$$

8. (a) Under what circumstances is the matrix A both symmetric and skew-symmetric?
   (b) Prove: If A is symmetric, then A^n is symmetric (n a positive integer).
   (c) If A is skew-symmetric, what can you say about A^n?

9. Let A, B be symmetric matrices of order n. Prove that AB is symmetric if and only if $AB = BA$.

10. The statement "$ax + by = 0$ for all x, y" is much stronger than the statement "$ax + by = 0$." The first statement implies that $a = b = 0$. Let A, B be matrices of type (m, n), and let $X = [x_1 \ \ldots \ x_n]$.
    (a) Prove: If $AX^T = 0$ for all X, then $A = 0$.
    (b) Prove: If $AX^T = BX^T$ for all X, then $A = B$.
    (c) Disprove: If $XAX^T = XBX^T$ for all X, then $A = B$.

11. Let XBX^T represent a quadratic form, with B symmetric. Explain why B is unique. You may use the fact that $b_{ij} + b_{ji}$ must be the total coefficient of $x_i x_j$, and that $b_{ij} = b_{ji}$.

12. Let A and B be two matrices of types (m, n) and (n, p). Show
    (a) the rows of AB are $A_1 B \ A_2 B, \ldots, A_m B$, where $A_1, A_2, \ldots, A_m$ are the rows of A.
    (b) the columns of AB are $AB_1, \ldots, AB_p$, where $B_1, \ldots, B_p$ are the columns of B.

13. Let A be any matrix of type (3, 4) and let
$$B = \begin{bmatrix} 0 & 1 & 0 \\ 1 & 0 & 0 \\ 0 & 0 & 1 \end{bmatrix}$$
    (a) How is A changed when it is premultiplied by B?
    (b) Find a matrix C so the columns of AC are the columns of A, in reverse order.

Some Uses of Matrix Notation and Matrix Algebra

Optional Exercises on Complex Matrices

14. Show that a real Hermitian matrix is symmetric, and that a real skew-Hermitian matrix is skew-symmetric.

15. Express the following matrix as the sum of two matrices B and C, with B Hermitian and C skew-Hermitian.
$$\begin{bmatrix} 2-i & 4+2i & 4 \\ -2i & 5 & 2+i \\ 2-2i & 6+5i & 4i \end{bmatrix}$$

16. Prove each of the following general laws:
    (a) $(A + B)^* = A^* + B^*$.
    (b) $(AB)^* = B^*A^*$.
    (c) $(A^*)^* = A$.
    (d) $(A^*)^n = (A^n)^*$ if n is a positive integer.
    (e) Theorem 10-12.

17. (a) Given
$$A = \begin{bmatrix} 1 & 2+i \\ 2-i & 2 \end{bmatrix},$$
a Hermitian matrix, write out the quadratic form XAX^*, and show that its value is real (for all complex X).
    (b) Prove: If A is Hermitian of order n, then XAX^* is real (for all complex X).

Remarks on Exercises

1. $\begin{bmatrix} 1 & 0 & 1 & 5 \\ 0 & 1 & 0 & 3 \\ 0 & 0 & 0 & 0 \end{bmatrix} \begin{bmatrix} x_1 \\ x_2 \\ x_3 \\ x_4 \end{bmatrix} = \begin{bmatrix} 3 \\ 1 \\ 2 \end{bmatrix}$. Inconsistent.

2. $x_1 + x_2 + 2x_3 = 0$, $x_1 + 2x_2 + 3x_3 = 0$, $2x_1 - x_2 + x_3 = 0$, $x_1 - 3x_2 - 2x_3 = 0$. The solution set is the set $\{t(1, 1, -1) \mid t \in R\}$.

3. (a) The equation $AX^T = B$ has a unique solution X, because the vector B is a unique linear combination of the basis consisting of the columns of A.
    (b) The equation $AX^T = 0$ has a nontrivial solution X.

4. (a) The definition of matrix multiplication requires that for each i, $\sum_{j=1}^{n} a_{ij}x_j = 0$, which means that $(a_{i1}, a_{i2}, \ldots, a_{in}) \cdot (x_1, \ldots, x_n) = 0$.

Lessons in Linear Algebra

(b) The plane $2x + y - z = 0$ is perpendicular to the vector $(2, 1, -1)$, because $(2, 1, -1) \cdot (x, y, z) = 0$ for all points (x, y, z) of the plane. A solution of both equations $2x + y - z = 0$ and $x + y - 5z = 0$ would be perpendicular to both of the vectors $(2, 1, -1)$ and $(1, 1, -5)$.

5. (a) $x_1^2 - x_2^2 + 5x_3^2 + 5x_1x_2 + 4x_1x_3 - 2x_2x_3$. At $(1, -1, 1)$, the value is 6.

(b) $B = \begin{bmatrix} 1 & \frac{5}{2} & 2 \\ \frac{5}{2} & -1 & -1 \\ 2 & -1 & 5 \end{bmatrix}$, $C = \begin{bmatrix} 0 & \frac{1}{2} & 5 \\ -\frac{1}{2} & 0 & 5 \\ -5 & -5 & 0 \end{bmatrix}$.

6. $\begin{bmatrix} 1 & 1 & -3 & 4 \\ 1 & 1 & 2 & 0 \\ -3 & 2 & 0 & -\frac{1}{2} \\ 4 & 0 & -\frac{1}{2} & -1 \end{bmatrix}$

7. (a) $XAX^T = x_1^2 + 2x_2^2 \geq 0$, etc.
   (b) $XBX^T \geq 0$ for all X. However, XBX^T can be 0 for a nonzero X. Try $X = (0, 0, 1)$.

8. (a) If and only if $A = 0$.
   (c) A^n is symmetric if n is even, skew-symmetric if n is odd.

10. (a) To say that $AX^T = 0$ for all X implies that each row of A is orthogonal to every vector X. Hence, each row of A is 0, for each row is self-orthogonal, for one thing.
    (b) If $(A - B)X^T = 0$ for all X, use (a).
    (c) There are many counterexamples. Let

$$A = \begin{bmatrix} 1 & 1 \\ 1 & 1 \end{bmatrix}, \quad B = \begin{bmatrix} 1 & 2 \\ 0 & 1 \end{bmatrix}$$

Then $XAX^T = XBX^T$ for all X, but $A \neq B$.

11. In the quadratic form XBX^T (B symmetric), the total coefficient c_{ij} of x_ix_j must equal $b_{ij} + b_{ji}$. But $b_{ij} = b_{ji}$, so $b_{ij} = (1/2)c_{ij}$, and hence, B is unique.

Some Uses of Matrix Notation and Matrix Algebra

12. (a) We simply use the definition of matrix multiplication.

    $$(\text{The } i\text{th row of } A) \, B = [a_{i1} \quad a_{i2} \quad \ldots \quad a_{in}] \begin{bmatrix} b_{11} & \cdots & b_{1p} \\ & \cdot & \\ & \cdot & \\ & \cdot & \\ b_{n1} & \cdots & b_{np} \end{bmatrix}$$

    $$= \left[\sum_{k=1}^{n} a_{ik}b_{k1}, \sum_{k=1}^{n} a_{ik}b_{k2}, \ldots, \sum_{k=1}^{n} a_{ik}b_{kp} \right] = (\text{the } i\text{th row of } AB)$$

13. (a) The first two rows of A are swapped.

    (b) $C = \begin{bmatrix} 0 & 0 & 0 & 1 \\ 0 & 0 & 1 & 0 \\ 0 & 1 & 0 & 0 \\ 1 & 0 & 0 & 0 \end{bmatrix}.$

14. If A is real $\bar{A} = A$ and $A^* = A^T$. Then a real Hermitian matrix A satisfies $A^* = A^T = A$.

15. $B = \begin{bmatrix} 2 & 2+2i & 3+i \\ 2-2i & 5 & 4-2i \\ 3-i & 4+2i & 0 \end{bmatrix}, C = \begin{bmatrix} -i & 2 & 1-i \\ -2 & 0 & -2+3i \\ -1-i & 2+3i & 4i \end{bmatrix}$

16. (b) $(AB)^* = \overline{AB}^T = (\bar{A}\bar{B})^T = (\bar{B})^T(\bar{A})^T = B^*A^*$.
    (d) Use mathematical induction. The statement holds for $n = 1$. Let $(A^*)^k = (A^k)^*$ and try to show that $(A^*)^{k+1} = (A^{k+1})^*$. Here are the steps: $(A^{k+1})^* = (A^k A)^* = A^*(A^k)^* = A^*(A^*)^k = (A^*)^{k+1}$.

17. (a) $XAX^* = x_1\bar{x}_1 + (2+i)x_1\bar{x}_2 + (2-i)x_2\bar{x}_1 + 2x_2\bar{x}_2$
    $= |x_1|^2 + (2+i)x_1\bar{x}_2 + \overline{(2+i)x_1\bar{x}_2} + 2|x_2|^2$
    $= |x_1|^2 + 2 \text{ Real } [(2+i)x_1\bar{x}_2] + 2|x_2|^2$.

    (b) For all complex X, we show that XAX^* is self-conjugate (therefore, real).

    $$\overline{XAX^*} = \overline{(XAX^*)}^T \quad (XAX^* \text{ is of type 1, 1.})$$
    $$= (XAX^*)^*$$
    $$= XA^*X^*$$
    $$= XAX^* \quad (A \text{ is Hermitian.})$$

11

ELEMENTARY MATRICES, THE INVERSE OF A MATRIX, AND RANK OF A MATRIX

In studying this lesson you will need to use the rules of matrix algebra. Also, you must recall some of the facts about the way in which the unique Hermite normal form of a matrix is used to solve the system $AX^T = B^T$ of linear equations.

The first item for discussion is the "inverse" of a matrix A. The product AB of the two matrices

$$A = \begin{bmatrix} 1 & 1 & 1 \\ 2 & 1 & 1 \end{bmatrix}, \quad B = \begin{bmatrix} -1 & 1 \\ 1 & -2 \\ 1 & 1 \end{bmatrix}$$

gives $AB = I$, the 2 by 2 identity matrix. However, BA is not an identity matrix. We say A is the "left-inverse" of B and B is the "right-inverse" of A. It is not difficult to show that if A is of type (m, n) with $m > n$, then A has no right-inverse (Exercise 5).

When we say A and B are inverses, we mean something much stronger.

11-1 Definition

Let A and B be square matrices with $AB = BA = I$. Then B is called the *inverse of A*, and we write $B = A^{-1}$. (Also, $A = B^{-1}$.) When A has an inverse we say A is "invertible" or "nonsingular." (A matrix which is not invertible is called "singular.")

The matrices

$$\begin{bmatrix} 1 & 1 \\ 2 & 3 \end{bmatrix} \text{ and } \begin{bmatrix} 3 & -1 \\ -2 & 1 \end{bmatrix}$$

are inverses. Not all square matrices have inverses as you can see by examining the matrix

$$\begin{bmatrix} 1 & 2 \\ 2 & 4 \end{bmatrix}$$

Elementary Matrices, the Inverse of a Matrix, and Rank of a Matrix

If the square matrix A has an inverse B, then B is unique.

11-2 Theorem

Proof: Let B, C be two inverses of A. This means of course that $BA = AB = CA = AC = I$. Premultiplying each member of $AC = I$ by B gives $BAC = BI$, $IC = B$, and $C = B$. ∎

Let A and B be invertible matrices. Then AB is invertible and $(AB)^{-1} = B^{-1}A^{-1}$.

11-3 Theorem

Proof: $B^{-1}A^{-1}$ is a right-inverse of AB, because $(AB)(B^{-1}A^{-1}) = AIA^{-1} = AA^{-1} = I$. Similarly, $B^{-1}A^{-1}$ is a left-inverse of AB. ∎

Before we proceed further with our study of inverses, we introduce a new and very simple kind of matrix, an "elementary" matrix. Elementary matrices are square matrices constructed very easily from an identity matrix.

From Definition 5-1, an *elementary operation* on a set $\{\alpha_1, \alpha_2, \ldots, \alpha_n\}$ is one of three things: (1) swapping two elements; (2) multiplying one element by a nonzero scalar; (3) adding to one element a multiple of another.

An *elementary* matrix is a matrix obtained by performing exactly one elementary operation on the rows (or columns) of an identity matrix.

11-4 Definition

Examples of elementary matrices are

$$E = \begin{bmatrix} 1 & 0 & 0 \\ 0 & 0 & 1 \\ 0 & 1 & 0 \end{bmatrix} \quad F = \begin{bmatrix} 1 & k & 0 & 0 \\ 0 & 1 & 0 & 0 \\ 0 & 0 & 1 & 0 \\ 0 & 0 & 0 & 1 \end{bmatrix} \quad G = \begin{bmatrix} 1 & 0 \\ 0 & 3 \end{bmatrix}$$

E is obtained by swapping rows 2 and 3 of the third-order identity I_3. F is obtained by starting with I_4 and adding k times the second row to the first row (or, if you prefer, by adding k times the first column to the second column). To get G we multiply the second row of I_2 by 3.

Observe that performing any elementary row operation on I is equivalent to performing an elementary column operation on I.

Now we consider what happens when we multiply a general matrix A by an elementary matrix E. Let

Lessons in Linear Algebra

$$E = \begin{bmatrix} 1 & 0 & 0 \\ 0 & 0 & 1 \\ 0 & 1 & 0 \end{bmatrix}, \quad A = \begin{bmatrix} a_{11} & a_{12} & a_{13} & a_{14} \\ a_{21} & a_{22} & a_{23} & a_{24} \\ a_{31} & a_{32} & a_{33} & a_{34} \end{bmatrix}$$

Then EA is the matrix A with *one* change. Rows 2 and 3 are swapped. But this is the same thing that was done to I to produce E.

Again, let

$$F = \begin{bmatrix} 1 & 2 & 0 \\ 0 & 1 & 0 \\ 0 & 0 & 1 \end{bmatrix}, \quad A = [a_{ij}]_{3,4}, \quad B = [b_{ij}]_{5,3}$$

Then FA is the matrix obtained from A by adding twice the second row to the first row. But this is how we obtain F from I. Also, BF is the matrix obtained from B by adding twice the first column to the second. But this is how we obtain F from I.

The generalization is simple and important.

11-5 Theorem

Let A be a matrix of type (m, n). Then any elementary operation on the rows (or columns) of A can be performed by multiplying A by an elementary matrix. Specifically,

(1) if E is an elementary matrix obtained by performing an elementary row operation on I_m, then EA is the matrix obtained by performing the same row operation on A.

(2) if F is an elementary matrix obtained by performing an elementary column operation of I_n, then AF is the matrix obtained by performing the same column operation on A.

Proof: The proof is simple but somewhat tedious. We take only one case of six. Let $A = [a_{ij}]_{m,n}$ and let E be obtained from the identity I_m by adding k times the ith row to the jth row. The ji-element of E is k; elsewhere E is the same as I_m. Letting $R_1, \ldots, R_m$ be the rows of A, we have

$$EA = \begin{bmatrix} 1 & \cdot & 0 & \cdot & 0 & \cdot & 0 \\ & & & & & & \\ & & \cdot & & \cdot & & \\ j & 0 & k & & 1 & & \\ & 0 & & \cdot & \cdot & \cdot & 1 \end{bmatrix} \begin{bmatrix} R_1 \\ \cdot \\ \cdot \\ \cdot \\ \cdot \\ \cdot \\ R_m \end{bmatrix} = \begin{bmatrix} R_1 \\ R_2 \\ \cdot \\ \cdot \\ kR_i + R_j \\ \cdot \\ R_m \end{bmatrix} \blacksquare$$

There are immediate consequences of importance.

Elementary Matrices, the Inverse of a Matrix, and Rank of a Matrix

Any matrix A can be reduced to Hermite normal form by premultiplying A by a finite number of elementary matrices.

11-6 Theorem

Proof: Each elementary row operation can be accomplished by premultiplication by an elementary matrix E_i. We get the Hermite normal form $E_r \ldots E_2 E_1 A$. ∎

Example 1. Let $A = \begin{bmatrix} 1 & 2 & 3 \\ -1 & 2 & 1 \end{bmatrix}$. Let $E_1 = \begin{bmatrix} 1 & 0 \\ 1 & 1 \end{bmatrix}$. Then $E_1 A = \begin{bmatrix} 1 & 2 & 3 \\ 0 & 4 & 4 \end{bmatrix}$. Let $E_2 = \begin{bmatrix} 1 & 0 \\ 0 & 1/4 \end{bmatrix}$. Then $E_2 E_1 A = \begin{bmatrix} 1 & 2 & 3 \\ 0 & 1 & 1 \end{bmatrix}$. Let $E_3 = \begin{bmatrix} 1 & -2 \\ 0 & 1 \end{bmatrix}$. Then $E_3 E_2 E_1 A = \begin{bmatrix} 1 & 0 & 1 \\ 0 & 1 & 1 \end{bmatrix}$.

The sequence E_i is not unique, because the sequence of elementary operations used to reduce A to Hermite normal form is not unique. The Hermite normal form itself is of course unique.

Next we comment on the invertibility of a square matrix.

Let A be an n by n square matrix, let $X = [x_1 \ldots x_n]$, $B = [b_1 \ldots b_n]$. The following four conditions are equivalent:

11-7 Theorem

(1) $AX^T = B^T$ has a unique solution X.
(2) The Hermite normal form of A is I.
(3) There exist elementary matrices $E_1, \ldots, E_r$ such that $E_r \ldots E_1 A = I$.
(4) A is invertible.

Proof: We show (1) $\Rightarrow$ (2) $\Rightarrow$ (3) $\Rightarrow$ (4) $\Rightarrow$ (1).

(1) $\Rightarrow$ (2): If the Hermite normal form of A were not I, then the last row of A would be 0. Then the system $AX^T = B$ would be either inconsistent or would have more than one solution.

(2) $\Rightarrow$ (3): This follows from Theorem 11-6.

We digress for a moment to show that any elementary matrix E has an inverse which is an elementary matrix. If E is obtained from I by an i, j row-swap, then E is its own inverse. If E is obtained by multiplying the ith row of I by k, then E^{-1} is obtained by multiplying the ith row of I by $1/k$. If E is obtained from I by adding k times the ith row to the jth row, then E^{-1} is obtained from I by adding $-k$ times the ith row to the jth row.

(3) $\Rightarrow$ (4): Let $E_1 \ldots E_r A = I$. Premultiply each member by $E_r^{-1} \ldots E_1^{-1}$ and get $A = E_r^{-1} \ldots E_1^{-1}$. Then post-multiplication by $E_1 \ldots E_r$ gives $AE_1 \ldots E_r = I$. Then we have $(E_1 \ldots E_r)A = A(E_1 \ldots E_r) = I$, and $A^{-1} = E_1 \ldots E_r$.

(4) $\Rightarrow$ (1): If A^{-1} exists, then $AX^T = B^T$ is equivalent to $X^T = A^{-1}B^T$, and X is unique. ∎

11-8 Theorem

If A is invertible then $(A^T)^{-1} = (A^{-1})^T$.

Proof: We use the definition of A^{-1} and the fact that $(MN)^T = N^T M^T$ for all matrices M, N.

$$(A^{-1})^T A^T = (AA^{-1})^T = I^T = I$$

This shows that $(A^{-1})^T$ is a left-inverse of A^T.

$$(A^T)(A^{-1})^T = (A^{-1}A)^T = I^T = I$$

This shows that $(A^{-1})^T$ is a right-inverse of A^T. ∎

11-9 Theorem

Let A be a square matrix. Then any one-sided (left or right) inverse of A is the inverse of A.

Proof: If A has a left-inverse C, then the system $AX^T = B^T$ has the unique solution $X^T = CB^T$. By Theorem 11-7, A is invertible. So we know that any left-inverse is the inverse.

If D is a right-inverse of A, then $AD = I$. Transposing gives $D^T A^T = I$ and D^T is a left-inverse of A^T, and as we have just shown, D^T is the inverse of A^T. Now, by Theorem 11-8, $D^T = (A^T)^{-1}$ implies $D^T = (A^{-1})^T$ and $D = A^{-1}$. ∎

We now consider the problem of inverting a square matrix. We know that it is enough to find a one-sided inverse. In some cases one can stumble upon A^{-1}. For example, suppose we are given that $A^2 + A + I = 0$. We use $I = A(-I - A)$ to detect that $A^{-1} = -I - A$.

A more general procedure arises out of the reduction of A to Hermite normal form $E_1 \ldots E_r A$, by means of the elementary matrices $E_1, \ldots, E_r$. If $E_1 \ldots E_r A = I$, then $E_1 \ldots E_r = A^{-1}$. If the Hermite normal form is not the identity, then A is not invertible. It is easy to keep track of things, at least when the matrices are small.

Elementary Matrices, the Inverse of a Matrix, and Rank of a Matrix

Example 2. Invert (if possible)

$$A = \begin{bmatrix} 2 & 1 \\ 1 & 1 \end{bmatrix}$$

Our plan is to reduce A to Hermite normal form by finding appropriate elementary matrices $E_1, E_2, \ldots$.

Pick $E_1 = \begin{bmatrix} 0 & 1 \\ 1 & 0 \end{bmatrix}$ $\qquad E_1 A = \begin{bmatrix} 1 & 1 \\ 2 & 1 \end{bmatrix}$

Let $E_2 = \begin{bmatrix} 1 & 0 \\ -2 & 1 \end{bmatrix}$ $\qquad E_2 E_1 A = \begin{bmatrix} 1 & 1 \\ 0 & -1 \end{bmatrix}$

Let $E_3 = \begin{bmatrix} 1 & 0 \\ 0 & -1 \end{bmatrix}$ $\qquad E_3 E_2 E_1 A = \begin{bmatrix} 1 & 1 \\ 0 & 1 \end{bmatrix}$

Let $E_4 = \begin{bmatrix} 1 & -1 \\ 0 & 1 \end{bmatrix}$ $\qquad E_4 E_3 E_2 E_1 A = I$

Therefore,

$$A^{-1} = E_4 E_3 E_2 E_1 = \begin{bmatrix} 1 & -1 \\ -1 & 2 \end{bmatrix}$$

We sketch the process again in order to be more efficient.

$$
\begin{array}{r|cc|cc|l}
I: & 1 & 0 & 2 & 1 & A \\
 & 0 & 1 & 1 & 1 & \\
\hline
E_1: & 0 & 1 & 1 & 1 & \\
 & 1 & 0 & 2 & 1 & \\
\hline
E_2 E_1: & 0 & 1 & 1 & 1 & \\
 & 1 & -2 & 0 & -1 & \\
\hline
E_3 E_2 E_1: & 0 & 1 & 1 & 1 & \\
 & -1 & 2 & 0 & 1 & \\
\hline
A^{-1}: & 1 & -1 & 1 & 0 & I \\
 & -1 & 2 & 0 & 1 & \\
\end{array}
$$

We start with I and A at the head of two columns. We apply elementary *row* operations to A until we get the Hermite normal form. At the same time we apply the identical row operations to the column headed by I. If the A column ends with I, then the other column ends with A^{-1}. If the Hermite normal form of A is not I, then A^{-1} does not exist.

Lessons in Linear Algebra

11-10 Theorem If A is invertible, then A is the product of elementary matrices.

The proof is left for you. (Hint: in Example 2, $A^{-1} = E_4 E_3 E_2 E_1$.)

We conclude this lesson with an introduction to the concept of *rank* of an m by n matrix. The subspace (of R^n) spanned by the rows of A is called the "row space" of A, and the subspace (of R^m) spanned by the columns of A is called the "column space" of A.

11-11 Theorem Let $A = [a_{ij}]_{m,n}$. The dimension of the row space of A is equal to the dimension of the column space of A.

Proof: First we show that applying an elementary row operation to A changes neither the dimension of the row space nor that of the column space. Let A and A' be such that either can be changed into the other by applying an elementary row operation.

Then each row of A' is a linear combination of the rows of A, and conversely. So the row spaces of A and A' are the same and hence have equal dimensions.

Now let us compare the column spaces of A and A'. Each linear relation among the columns of A leads to a linear relation among the same columns of A'. For example, compare the linear relations

$$c_1 \begin{bmatrix} a_{11} \\ a_{21} \\ \vdots \end{bmatrix} + c_2 \begin{bmatrix} a_{12} \\ a_{22} \\ \vdots \end{bmatrix} + \ldots = 0$$

$$c_1 \begin{bmatrix} a_{11} + k a_{21} \\ a_{21} \\ \vdots \end{bmatrix} + c_2 \begin{bmatrix} a_{12} + k a_{22} \\ a_{22} \\ \vdots \end{bmatrix} + \ldots = 0$$

If one holds, so does the other. This means that any linearly independent set of columns in A corresponds to the linearly independent set of the same columns in A'. Therefore, the dimensions of the two column spaces are equal.

Now let B be the Hermite normal form of A. Since we can get from A to B by performing elementary row operations, the two matrices have row spaces of the same dimension r and column spaces of the same dimension c.

Elementary Matrices, the Inverse of a Matrix, and Rank of a Matrix

A close look at the Hermite normal form should convince you that $r \leq c$. Appy this principle to A^T and we have $c \leq r$. Therefore, $r = c$. That is, the dimensions of the row space and column space of A are equal. ∎

The *rank* of a matrix is the dimension of its row space (or column space).

11-12
Definition

Example 3. Let

$$A = \begin{bmatrix} 1 & 2 & -1 & 1 \\ 3 & 7 & -3 & 2 \\ 1 & 1 & -1 & 2 \end{bmatrix}$$

The Hermite normal form of A is

$$A' = \begin{bmatrix} 1 & 0 & -1 & 3 \\ 0 & 1 & 0 & -1 \\ 0 & 0 & 0 & 0 \end{bmatrix}$$

The rank of A is 2, the number of nonzero rows in A. This means A has two linearly independent rows and two linearly independent columns.

The rank of the product AB of two matrices cannot exceed the rank of either matrix.

11-13
Theorem

Proof: Rank $(AB) \leq$ rank B because each row of AB is a linear combination of the rows of B, and hence the row space of AB is a subspace of the row space of B. Also, rank $(AB) =$ rank $(AB)^T =$ rank $(B^T A^T) \leq$ rank $A^T =$ rank A.

Summary of Topics in Lesson 11

Definition of inverse of a matrix.

Uniqueness of the inverse matrix.

Elementary matrices and their use in performing elementary operations on other matrices.

The reduction of a matrix A to its Hermite normal form $E_r \ldots E_1 A$ by means of elementary matrices $E_1, \ldots, E_r$.

How to determine if a matrix is invertible, and how to find the inverse of an invertible matrix.

Lessons in Linear Algebra

Conditions that a square matrix be invertible.
An invertible matrix is the product of elementary matrices.
The rank of a matrix.

Exercises

1. Show that $\begin{bmatrix} 1 & 2 \\ 2 & 4 \end{bmatrix}$ has no inverse.

2. Invert (find the inverse of) each of the following:

   (a) $\begin{bmatrix} 3 & 4 \\ -5 & 6 \end{bmatrix}.$

   (b) $\begin{bmatrix} 1 & 1 & -1 \\ 2 & 5 & -1 \\ 2 & 0 & -3 \end{bmatrix}.$

   (c) $\begin{bmatrix} 1 & 1 & 2 & 1 \\ 0 & 1 & -1 & 4 \\ 1 & 2 & 3 & 1 \\ 0 & 0 & 0 & 2 \end{bmatrix}.$

   (d) $\begin{bmatrix} 1 & 2 & -1 \\ 3 & 1 & -4 \\ 1 & -3 & -2 \end{bmatrix}.$

3. (a) Verify: If $A = \begin{bmatrix} a & b \\ c & d \end{bmatrix}$ and $ad \neq bc$, then
   $$A^{-1} = \frac{1}{ad - bc} \begin{bmatrix} d & -b \\ -c & a \end{bmatrix}.$$

   (b) Use the rule (a) to invert $A = \begin{bmatrix} 2 & 3 \\ 1 & 5 \end{bmatrix}.$

   (c) Write out the following system in matrix notation, and use (b) to solve it.
   $$2x + 3y = 7$$
   $$x + 5y = -2$$

   (d) Let $A = \begin{bmatrix} 1 & 1 & -1 \\ 2 & 5 & -1 \\ 2 & 0 & -3 \end{bmatrix}$, the matrix of exercise 2(b). Write out and solve the system $AX^T = B^T$, where $B = [1 \quad 3 \quad 4].$

Elementary Matrices, the Inverse of a Matrix, and Rank of a Matrix

4. Let $A = \begin{bmatrix} 1 & 3 & 5 & -1 \\ -1 & -2 & 1 & 4 \\ 2 & 7 & 16 & 1 \end{bmatrix}$, $X = [x_1 \ x_2 \ x_3 \ x_4]$, and $B = [1 \ 0 \ 3]$.

   (a) Find a matrix C such that CA is the Hermite normal form of A.
   (b) Solve $AX^T = B^T$ by multiplying by C.
   (c) What is the rank of A?

5. Let A be an m by n matrix.
   (a) Show that A has no right-inverse, if $m > n$. (Hint: Each row of AB is a linear combination of the rows of B.)
   (b) Use (a) to argue that if $m < n$, then A has no left-inverse.

6. Prove or disprove: If A and B are invertible, then $(A + B)^{-1} = A^{-1} + B^{-1}$.

7. Let
$$A = \begin{bmatrix} 1 & 3 & 0 & 0 & 3 & 2 & 1 \\ 0 & 0 & 1 & 0 & -1 & 1 & 4 \\ 0 & 0 & 0 & 1 & 1 & 3 & 6 \\ 0 & 0 & 0 & 0 & 0 & 0 & 0 \end{bmatrix}$$

   (a) What is the rank of A?
   (b) Find a basis for the column space of A.

8. Given that $A^3 - I = 0$ and that $A - I$ is invertible, show that $A^{-1} = -A - I$.

9. The *trace* of a square matrix A, denoted $Tr(A)$, is the sum of the diagonal elements of A, that is, $Tr(A) = \sum_{i=1}^{n} a_{ii}$.

   (a) For $A = \begin{bmatrix} 1 & 2 \\ 2 & -1 \end{bmatrix}$ and $B = \begin{bmatrix} 3 & 4 \\ -5 & 1 \end{bmatrix}$ verify that $Tr(AB) = Tr(BA)$.
   (b) Prove: If A and B are n by n matrices, then $Tr(AB) = Tr(BA)$, $Tr(A + B) = Tr(A) + Tr(B)$.
   (c) Prove or disprove: If A is of type (m, n) and B is of type (n, m), then $Tr(AB) = Tr(BA)$.
   (d) Show that it is impossible to find two matrices A and B such that $AB - BA = I$.

10. Let A be an n by n matrix. Prove:
    (a) A is invertible if and only if rank $(A) = n$.

Lessons in Linear Algebra

(b) A is invertible if and only if A has n linearly independent rows (columns).

11. Let A be any m by n matrix. Show that there exists a nonsingular matrix P such that PA is the Hermite normal form of A.

12. Let A be an n by n matrix, and let $X = [x_1 \ldots x_n]$ and $B = [b_1 \ldots b_n]$. Prove that $XA = B$ has a unique solution X if and only if A is invertible.

13. The English alphabet is symbolized 1, 2, ..., 26 and a 2 by 4 message matrix is encoded by multiplying it by $\begin{bmatrix} 1 & 2 \\ 2 & 5 \end{bmatrix}$. Decode the message received, namely $\begin{bmatrix} 33 & 35 & 36 & 20 \\ 73 & 85 & 84 & 44 \end{bmatrix}$.

Optional Exercises on Complex Matrices

14. Find the inverse of $\begin{bmatrix} 2+i & 1-i \\ 3-i & 2 \end{bmatrix}$.

15. Let $A = \begin{bmatrix} 1 & -i & 1 & 0 \\ 2 & 3i & 1 & 1+i \end{bmatrix}$. Find a nonsingular matrix P such that PA is in Hermite normal form.

16. Given that A is invertible, prove:
    (a) $(\overline{A})^{-1} = \overline{A^{-1}}$.
    (b) $(A^*)^{-1} = (A^{-1})^*$.

Remarks on Exercises

2. (a) $\dfrac{1}{38}\begin{bmatrix} 6 & -4 \\ 5 & 3 \end{bmatrix}$.

   (b) $\begin{bmatrix} 15 & -3 & -4 \\ -4 & 1 & 1 \\ 10 & -2 & -3 \end{bmatrix}$.

   (c) $\dfrac{1}{2}\begin{bmatrix} 5 & 1 & -3 & -3 \\ -1 & 1 & 1 & -2 \\ -1 & -1 & 1 & 2 \\ 0 & 0 & 0 & 1 \end{bmatrix}$.

   (d) Not invertible.

Elementary Matrices, the Inverse of a Matrix, and Rank of a Matrix

3. (c) $AX^T = B^T$, where $X = [x \ y]$, $B = [7 \ -2]$.

   $X^T = A^{-1}B^T = \frac{1}{7}\begin{bmatrix} 5 & -3 \\ -1 & 2 \end{bmatrix}\begin{bmatrix} 7 \\ -2 \end{bmatrix} = \frac{1}{7}\begin{bmatrix} 41 \\ -11 \end{bmatrix}$.

   (d) $X = [-10 \ 3 \ -8]$.

4. (a) $C = \begin{bmatrix} -2 & -3 & 0 \\ 1 & 1 & 0 \\ -3 & -1 & 1 \end{bmatrix}$.

   (b) $\{(-2, 1, 0, 0) + x_3(13, -6, 1, 0) + x_4(10, -3, 0, 1) \mid x_3, x_4$ arbitrary$\}$.

   (c) 2.

5. (a) Suppose $AB = I_m$. The row space S of AB is spanned by the n rows of B, and so the dimension of S is at most n. Therefore, we have contradicted the assumption that $AB = I_m$, because the row space of I_m has dimension $m > n$.

   (b) "Transpose" the result in (a).

6. The statement $(A + B)^{-1} = A^{-1} + B^{-1}$ is not generally true. A counterexample is $A = I_2$, $B = I_2$. Then $(A + B)^{-1} = (2I)^{-1} = (1/2)I$, but $A^{-1} + B^{-1} = 2I_2$. In fact $A + B$ may not even be invertible, even though A and B are invertible. A more difficult problem is to find some pair A, B for which $(A + B)^{-1} = A^{-1} + B^{-1}$.

7. (a) 3.
   (b) The transposes of $\{(1, 0, 0, 0), (0, 1, 0, 0), (0, 0, 1, 0)\}$.

8. Factor $A^3 - I$.

9. (b) $Tr(AB) = \sum_{i=1}^{n}$ (the ii-element of AB)

   $= \sum_{i=1}^{n}\sum_{k=1}^{n} a_{ik}b_{ki}$.

   On the other hand,

   $Tr(BA) = \sum_{k=1}^{n}$ (the kk-element of BA)

   $= \sum_{k=1}^{n}\sum_{i=1}^{n} b_{ki}a_{ik}$

   (c) The proof parallels that of part (b).
   (d) Compare the traces of $AB - BA$ and I.

Lessons in Linear Algebra

10. (a) By Theorem 11-7, A is invertible if and only if there exist elementary matrices $E_1, \ldots, E_r$ such that $E_r \ldots E_1 A = I$. Then rank $(A) =$ rank $(I) = n$, because multiplying a matrix by an elementary matrix does not change the rank.
    (b) The rank of $A = n$ if and only if the row space of A has dimension n, or if and only if A has n linearly independent rows. Now use (a).

11. Let $E_r \ldots E_1 P$ be the Hermite normal form of A, with $E_1, \ldots, E_r$ elementary matrices. Then take $P = E_r \ldots E_1$.

12. $XA = B$ is equivalent to $A^T X^T = B^T$, which by Theorem 11-7 has a unique solution.

13. SELL GOLD.

14. $\dfrac{1}{20} \begin{bmatrix} 2 - 6i & 2 + 4i \\ 10i & 5 - 5i \end{bmatrix}$.

15. $P = \dfrac{1}{5} \begin{bmatrix} 3 & 1 \\ 2i & -i \end{bmatrix}$. $PA = \begin{bmatrix} 1 & 0 & \frac{4}{5} & \frac{1}{5}(1+i) \\ 0 & 1 & \frac{1}{5}i & \frac{1}{5}(1-i) \end{bmatrix}$.

16. (a) $(\overline{A^{-1}})(\overline{A}) = \overline{A^{-1}A} = \overline{I} = I$. Hence $\overline{A^{-1}} = (\overline{A})^{-1}$.
    (b) $(A^{-1})^* A^* = (AA^{-1})^* = I^* = I$. Hence $(A^{-1})^* = (A^*)^{-1}$.

12

LINEAR TRANSFORMATIONS

In this lesson you are introduced to very important mappings called "linear mappings" or "linear transformations." Actually you have used such mappings in your study of elementary algebra, geometry, and calculus. For example, the process of differentiation is linear.

After defining a linear transformation, we give several examples of such transformations and discuss some of their general properties. Then we explain how any linear transformation can be represented by a matrix, so that in the future we may treat the algebra of linear transformations as an algebra of matrices. It is important that you understand the equivalence of these two "linear" algebras.

You may find it profitable to review some of the technical terms of mappings. The summary in the preliminary lesson includes a discussion of general mappings. Of course, any property that holds for all mappings will hold for linear mappings. A linear transformation is a very special mapping, special in the sense that linearity is a strong assumption. So you would expect many results for linear mappings in addition to those which hold for all mappings.

Let V and W be two linear spaces over the same field F (See Figure 12-1). The mapping $T : V \to W$ is called a *linear transformation of V into W* if

$$T(a\alpha + b\beta) = aT(\alpha) + bT(\beta) \quad \text{for all } \alpha, \beta \in V, a, b \in F$$

12-1 Definition

12-2 Linearity Condition

The "linearity" condition (12-2) includes the quantification that it must hold for *all* vectors α, β and *all* scalars a, b.

Example 1. Let $T : R^3 \to R^2$ according to the rule $T(x, y, z) = (x - y, x + z)$. This is clearly a mapping of R^3 into R^2, because T assigns to each element (x, y, z) of R^3 a unique image $(x - y, x + z)$ of R^2. To illustrate $T(1, 2, 3) = (-1, 4)$. Also, $(3, 4, 5)$ goes into $(-1, 8)$.

Lessons in Linear Algebra

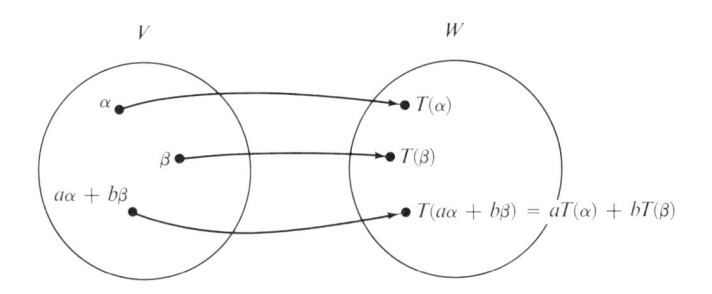

FIGURE 12-1

To show that T is linear we must verify that the linearity condition (12-2) holds for all $\alpha, \beta \in R^3$ and $a, b \in R$.

Let $\alpha = (x_1, x_2, x_3)$, $\beta = (y_1, y_2, y_3)$ be typical vectors of R^3, and let $a, b \in R$. Then

$$T(a\alpha + b\beta) = T[a(x_1, x_2, x_3) + b(y_1, y_2, y_3)]$$
$$= T(ax_1 + by_1, ax_2 + by_2, ax_3 + by_3)$$
$$= (ax_1 + by_1 - ax_2 - by_2, ax_1 + by_1 + ax_3 + by_3)$$

(by the rule for T)

On the other hand,

$$aT(\alpha) + bT(\beta) = a(x_1 - x_2, x_1 + x_3) + b(y_1 - y_2, y_1 + y_3)$$
$$= (ax_1 - ax_2 + by_1 - by_2, ax_1 + ax_3 + by_1 + by_3)$$

Therefore, $T(a\alpha + b\beta) = aT(\alpha) + bT(\beta)$ for all $\alpha, \beta \in R^3$, $a, b \in R$.

Example 2. Let V be the space of real polynomials $p(x)$ and define $D: V \to V$ by

$$D(p(x)) = p'(x)$$

Then D is linear because for all polynomials $p(x), q(x)$, and scalars a, b

$$D[ap + bq] = (ap + bq)'$$
$$= ap' + bq'$$
$$= aD(p) + bD(q)$$

Example 3. Let $T: R^2 \to R^2$ according to $T(x, y) = (x, x + y^2)$. This is a mapping but is *not* linear. For one thing, let $\alpha = (1, 1)$, $\beta = (1, 2)$. Then $T(\alpha + \beta) \neq T(\alpha) + T(\beta)$, because $(2, 11) \neq (1, 2) + (1, 5)$.

Linear Transformations

To show that a mapping is not linear a single counterexample suffices. To show linearity you must prove $T(a\alpha + b\beta) = aT(\alpha) + bT(\beta)$ for *all* α, β, a, b.

The mapping $f: A \to B$ is said to be a *one-to-one* mapping of A into B if whenever $a_1, a_2 \in A$ and $a_1 \neq a_2$, then $f(a_1) \neq f(a_2)$, or equivalently, if $f(a_1) = f(a_2)$ implies $a_1 = a_2$.

The mapping $f: A \to B$ is said to be a mapping of A *onto* B if each element of B is the image of at least one element of A, or if $b \in B$ implies $b = f(a)$ where $a \in A$.

Example 4. Let $T: R^3 \to R^2$ by $T(x, y, z) = (x, 2y)$. Then T is not 1-1, because the vectors $(1, 1, 3)$ and $(1, 1, 4)$ have the same image $(1, 2)$. However, T is a mapping of R^3 onto R^2 because the general vector (x, y) of R^2 is the image of $(x, (1/2)y, z)$.

Now, let $S: R^2 \to R^3$ by $S(x, y) = (x, y, x + y)$. Then S is 1-1 because $(x_1, y_1) \neq (x_2, y_2)$ implies $S(x_1, y_1) \neq S(x_2, y_2)$. However, S is not a mapping of R^2 *onto* R^3, because there is a vector in R^3 which is not the image of any vector of R^2. Take $(1, 2, 5)$. This is not of the form $(x, y, x + y)$ so it cannot be the image of any vector.

You should construct simple examples which show that a mapping may be both 1-1 and onto, or may be neither 1-1 nor onto.

We now turn to some of the special properties of linear transformations. We begin by giving an equivalent to the linearity condition (12-2).

12-3 Theorem

The mapping $T: V \to W$ is linear if and only if the following two conditions hold:

(1) $T(\alpha + \beta) = T(\alpha) + T(\beta)$, for all $\alpha, \beta \in V$.
(2) $T(a\alpha) = aT(\alpha)$, for all $\alpha \in V$, $a \in F$.

Proof: If $T(a\alpha + b\beta) = aT(\alpha) + bT(\beta)$ for all a, b, α, β, then special choices of a, b, α, β give (1) and (2). Take $a = b = 1$. Then take $b = 0$.

Conversely, suppose we have (1) and (2). Then for $a, b \in F$ and $\alpha, \beta \in V$ we have $T(a\alpha) = aT(\alpha)$, $T(b\beta) = bT(\beta)$ by (2), and $T(a\alpha + b\beta) = T(a\alpha) + T(b\beta)$ by (1). This yields the linearity condition $T(a\alpha + b\beta) = aT(\alpha) + bT(\beta)$. ∎

Now we speak of the linearity condition either in the form (12-2) or in the form (1) and (2) of Theorem 12-3.

Let T be a mapping of V into W. If we require that T be linear, then T is completely determined by its action on any basis of V.

Lessons in Linear Algebra

12-4 Theorem

Let $\{\alpha_1, \ldots, \alpha_n\}$ be a basis of V and let $\{\beta_1, \ldots, \beta_n\}$ be any vectors in W. There is exactly one linear transformation $T: V \to W$ for which $T(\alpha_i) = \beta_i$, $i = 1, 2, \ldots, n$.

Proof: The general vector α of V can be written $\alpha = \sum_{i=1}^{n} x_i \alpha_i$. Define $T(\alpha) = \sum_{i=1}^{n} x_i \beta_i$. Then T is linear because for all $a, a' \varepsilon F$, $\alpha, \alpha' \varepsilon V$.

$$T(a\alpha + a'\alpha') = T(a \sum x_i \alpha_i + a' \sum x'_i \alpha_i)$$
$$= T\left(\sum (ax_i + a'x'_i)\alpha_i\right)$$
$$= \sum (ax_i + a'x'_i)\beta_i \quad \text{(definition of } T\text{)}$$
$$= a \sum x_i \beta_i + a' \sum x'_i \beta_i$$
$$= aT(\alpha) + a'T(\alpha')$$

Now we show uniqueness by proving that any two linear transformations which agree on a basis agree everywhere. Let S, T be two linear transformations such that $S(\alpha_i) = T(\alpha_i)$ for $i = 1, \ldots, n$. Then $S(\sum x_i \alpha_i) = \sum x_i S(\alpha_i) = \sum x_i T(\alpha_i) = T(\sum x_i \alpha_i)$, because of linearity. Hence, S and T are equal on V, that is, $S(\alpha) = T(\alpha)$ for all $\alpha \varepsilon V$. ∎

Example 5. Let T be a linear transformation of R^3 into R^2 such that $T(1, 0, 0) = (1, 2)$, $T(0, 1, 0) = (3, -2)$ and $T(0, 0, 1) = (1, -1)$. Since we know the images of each element in a basis of R^3, the linear transformation is completely determined. It is easy to determine the image of a general vector.

$$(x, y, z) = x(1, 0, 0) + y(0, 1, 0) + z(0, 1, 0)$$

Therefore,

$$T(x, y, z) = xT(1, 0, 0) + yT(0, 1, 0) + zT(0, 0, 1) \quad \text{(linearity)}$$
$$= x(1, 2) + y(3, -2) + z(1, -1)$$
$$= (x + 3y + z, 2x - 2y - z)$$

If V and W are finite dimensional spaces with dimensions n and m respectively, then each transformation T can be represented by an m by n matrix.

12-5 Definition

Let T be a linear transformation of V into W and let $\{\alpha_1, \ldots, \alpha_n\}$, $\{\beta_1, \ldots, \beta_m\}$ be bases of V and W, respectively. The *matrix of T relative to the bases* $\{\alpha_i\}$ and $\{\beta_i\}$ is $A = [a_{ij}]_{m,n}$, where a_{ij} is defined by

$$T(\alpha_j) = \sum_{i=1}^{m} a_{ij}\beta_i, \quad j = 1, 2, \ldots, n$$

You may remember the above by noticing that we express the images of the α_i's in terms of the β_i's.

Example 6. Let $T: R^3 \to R^2$ by $T(x, y, z) = (x + y, y - z)$, and choose the bases $\{(0, 1, 1), (1, 0, 1), (1, 1, 0)\}$ and $\{(1, 1), (1, -1)\}$ of R^3 and R^2, respectively. In order to find the matrix of T relative to the two given bases, we express $T(0, 1, 1)$, $T(1, 0, 1)$, $T(1, 1, 0)$ in terms of $(1, 1)$, $(1, -1)$.

$$T(0, 1, 1) = (1, 0) = \frac{1}{2}(1, 1) + \frac{1}{2}(1, -1)$$

$$T(1, 0, 1) = (1, -1) = 0(1, 1) + 1(1, -1)$$

$$T(1, 1, 0) = (2, 1) = \frac{3}{2}(1, 1) + \frac{1}{2}(1, -1)$$

The matrix of T relative to the given bases is, therefore,

$$A = \begin{bmatrix} \frac{1}{2} & 0 & \frac{3}{2} \\ \frac{1}{2} & 1 & \frac{1}{2} \end{bmatrix}$$

In mechanically writing down the matrix we transposed the array as it appeared above. Actually we might have chosen A^T as the matrix representing T. The important thing is to make a choice as to which matrix will represent T and to use this choice consistently.

Remark: If T is a linear transformation of a space V into itself, then the "matrix of T relative to a given basis" is found by using the same basis in both the domain and codomain of T. That is the matrix of T relative to the basis $\{\alpha_1, \ldots, \alpha_n\}$ is $A = [a_{ij}]_{n,n}$, where

$$T(\alpha_j) = \sum_{i=1}^{n} a_{ij}\alpha_i, \quad j = 1, 2, \ldots, n$$

This amounts to setting $m = n$ and $\beta_i = \alpha_i$ in Definition 12-5.

Example 7. Let V be the space of real polynomials of degree less than or equal to 2, and let the linear transformation $T: V \to V$ be given by

$T(p(x)) = 3p''(x) - p'(x) + 2p(x)$. Let us find the matrix of T relative to the basis $\{1, x, x^2\}$.

Our problem is to express the images of the basis elements in terms of the basis.

$$T(1) = 2 \qquad\qquad = 2(1) + 0(x) + 0(x^2)$$
$$T(x) = -1 + 2x \qquad = -1(1) + 2(x) + 0(x^2)$$
$$T(x^2) = 6 - 2x + 2x^2 = 6(1) - 2(x) + 2x^2$$

The required matrix is

$$\begin{bmatrix} 2 & -1 & 6 \\ 0 & 2 & -2 \\ 0 & 0 & 2 \end{bmatrix}$$

We conclude this lesson with some remarks about the composite of two linear transformations. It will be our first step in showing how the algebra of linear transformations can be identified with the algebra of matrices. First we give two examples, then a generalization.

Example 8. Let $T: R^2 \to R^4$ and $S: R^4 \to R^3$ be the linear transformations defined by

$$T(x_1, x_2) = (x_1, 2x_2, x_1 + x_2, x_1 - x_2)$$
$$S(x_1, x_2, x_3, x_4) = (x_1 + x_2, -x_3, x_2 + x_4)$$

(See Figure 12-2.) Then ST is a linear transformation of $R^2 \to R^3$, with

$$ST(x_1, x_2) = S(x_1, 2x_2, x_1 + x_2, x_1 - x_2)$$

or

$$ST(x_1, x_2) = (x_1 + 2x_2, -x_1 - x_2, x_1 + x_2)$$

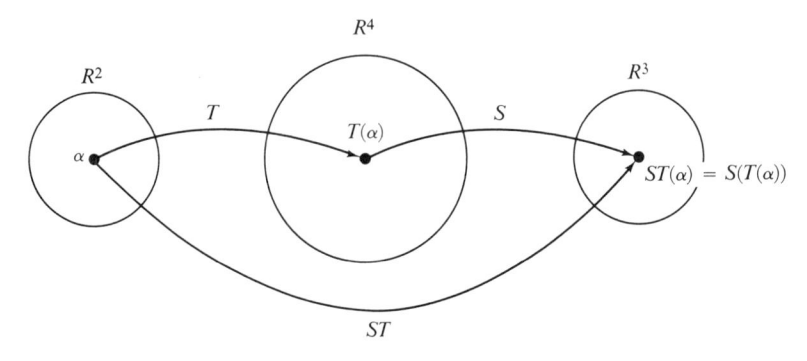

FIGURE 12-2

Linear Transformations

Now we represent each of the three transformations T, S, ST as matrices. To this purpose we choose bases for each of the three spaces.

$\alpha = \{\alpha_1, \alpha_2\}$ where $\alpha_1 = (1, 0)$, $\alpha_2 = (0, 1)$
$\mathcal{B} = \{\beta_1, \beta_2, \beta_3, \beta_4\}$ where $\beta_1 = (1, 0, 0, 0)$, $\beta_2 = (0, 1, 0, 0)$, etc.
$\mathcal{C} = \{\gamma_1, \gamma_2, \gamma_3\}$ where $\gamma_1 = (1, 0, 0)$, $\gamma_2 = (0, 1, 0)$, $\gamma_3 = (0, 0, 1)$

We have selected the "standard" bases for each of the three spaces so as to make the computation easier.

To find the matrix of T relative to the bases α and $\mathcal{B}$, we find the T-images of the elements of α in terms of the elements of $\mathcal{B}$.

$$T(1, 0) = (1, 0, 1, 1) = 1\beta_1 + 0\beta_2 + 1\beta_3 + 1\beta_4$$
$$T(0, 1) = (0, 2, 1, -1) = 0\beta_1 + 2\beta_2 + 1\beta_3 - 1\beta_4$$

Therefore, the matrix of T relative to α, $\mathcal{B}$ is

$$A = \begin{bmatrix} 1 & 0 \\ 0 & 2 \\ 1 & 1 \\ 1 & -1 \end{bmatrix}$$

The choice of a standard basis makes the columns of A very easy to find.

In the same way we find the matrix of S relative to the bases $\mathcal{B}$, $\mathcal{C}$.

$$S(1, 0, 0, 0) = (1, 0, 0) \quad = \gamma_1$$
$$S(0, 1, 0, 0) = (1, 0, 1) \quad = \gamma_1 \quad\quad + \gamma_3$$
$$S(0, 0, 1, 0) = (0, -1, 0) = \quad\quad -\gamma_2$$
$$S(0, 0, 0, 1) = (0, 0, 1) \quad = \quad\quad\quad\quad\quad \gamma_3$$

Therefore, the matrix of S relative to $\mathcal{B}$, $\mathcal{C}$ is

$$B = \begin{bmatrix} 1 & 1 & 0 & 0 \\ 0 & 0 & -1 & 0 \\ 0 & 1 & 0 & 1 \end{bmatrix}$$

Finally,

$$ST(1, 0) = S(1, 0, 1, 1) \quad = (1, -1, 1) = \gamma_1 - \gamma_2 + \gamma_3$$
$$ST(0, 1) = S(0, 2, 1, -1) = (2, -1, 1) = 2\gamma_1 - \gamma_2 + \gamma_3$$

and the matrix of ST relative to the bases α, $\mathcal{C}$ is

$$C = \begin{bmatrix} 1 & 2 \\ -1 & -1 \\ 1 & 1 \end{bmatrix}$$

Lessons in Linear Algebra

Note that $C = BA$, that is, the *matrix of ST is equal to the matrix of S multiplied by the matrix of T*. That the composition of linear transformations can be represented by matrix multiplication is general, as you will see in Theorem 12-6.

Example 9. If the bases are not the standard ones the arithmetic may be harder, but we can still represent linear transformations by matrices. Let S and T be linear transformations of R^2 into R^2, with

$$S(x, y) = (x + 2y, 3x)$$
$$T(x, y) = (4x, x + y)$$

We represent the transformations T, S, ST relative to the basis $\{\alpha_1, \alpha_2\}$ where $\alpha_1 = (1, 1)$, $\alpha_2 = (1, -1)$.

$$T(1, 1) = (4, 2) = 3\alpha_1 + \alpha_2$$
$$T(1, -1) = (4, 0) = 2\alpha_1 + 2\alpha_2$$

and the matrix of T relative to the given basis is

$$A = \begin{bmatrix} 3 & 2 \\ 1 & 2 \end{bmatrix}$$

Next

$$S(1, 1) = (3, 3) = 3\alpha_1$$
$$S(1, -1) = (-1, 3) = \alpha_1 - 2\alpha_2$$

and the matrix of S relative to the given basis is

$$B = \begin{bmatrix} 3 & 1 \\ 0 & -2 \end{bmatrix}$$

Also,

$$ST(1, 1) = S(4, 2) = (8, 12) = 10\alpha_1 - 2\alpha_2$$
$$ST(1, -1) = S(4, 0) = (4, 12) = 8\alpha_1 - 4\alpha_2$$

and the matrix of ST relative to the given basis is

$$C = \begin{bmatrix} 10 & 8 \\ -2 & -4 \end{bmatrix}$$

Again notice that $C = BA$, or relative to the given basis,

matrix of ST = (matrix of S)(matrix of T)

Linear Transformations

12-6 Theorem

Let U, V, W be finite-dimensional linear spaces with bases $\alpha = \{\alpha_1, \ldots, \alpha_n\}$, $\mathcal{B} = \{\beta_1, \cdots, \beta_m\}$, $\mathcal{C} = \{\gamma_1, \cdots, \gamma_p\}$, respectively. Let T be a linear transformation of U into V, and let S be a linear transformation of V into W. Then

(1) ST is a linear transformation of U into W.
(2) if A, B, C are the matrices of T, S, ST, respectively, relative to the given bases, then $C = BA$.

Proof: (1) Let α, β be any two vectors of U, and let a, b be scalars.

$$\begin{aligned}
ST(a\alpha + b\beta) &= S(T(a\alpha + b\beta)) && \text{(composition of mappings)} \\
&= S(aT(\alpha) + bT(\beta)) && \text{(linearity of } T) \\
&= aS(T(\alpha)) + bS(T(\beta)) && \text{(linearity of } S) \\
&= aST(\alpha) + bST(\beta) && \text{(composition of mappings)}
\end{aligned}$$

Therefore, ST is linear.

(2) By Definition 12-5 of matrix of a linear transformation, we can determine the matrices A, B, C of the transformations T, S, ST.

Let $A = [a_{ij}]_{m,n}$, $B = [b_{ij}]_{p,m}$, $C = [c_{ij}]_{p,n}$. For $j = 1, \ldots, n$, we have

$$\begin{aligned}
ST(\alpha_j) &= S\left(\sum_{k=1}^{m} a_{kj}\beta_k\right) && \text{(Definition 12-5)} \\
&= \sum_{k=1}^{m} a_{kj} S(\beta_k) && \text{(linearity of } S) \\
&= \sum_{k=1}^{m} a_{kj} \sum_{i=1}^{p} b_{ik}\gamma_i && \text{(Definition 12-5)} \\
&= \sum_{i=1}^{p} \sum_{k=1}^{m} b_{ik} a_{kj} \gamma_i && \text{(properties of summations)}
\end{aligned}$$

But by Definition 12-5,

$$ST(\alpha_j) = \sum_{i=1}^{p} c_{ij}\gamma_i$$

Therefore, $c_{ij} = \sum_{k=1}^{m} b_{ik}a_{kj}$, or $C = BA$. ∎

Summary of Topics in Lesson 12

Definition of linear transformation of one vector space into another.

Lessons in Linear Algebra

Examples of linear transformations.

How to determine if a given mapping is linear.

A linear transformation is determined uniquely by its action on a basis.

The matrix of a linear transformation, relative to given bases.

Relative to given bases, the matrix of the product (composite) of two linear transformations is the product of their matrices.

Exercises

1. Which of the following mappings of $R^3 \to R^2$ are linear? 1-1? onto?

   (a) $T(x, y, z) = (3x + y, x - 2y + z)$
   (b) $T(x, y, z) = (x, x + y + z + 1)$
   (c) $T(x, y, z) = (x + y - z, 2x + 2y - 2z)$
   (d) $T(x, y, z) = (0, 0)$
   (e) $T(x, y, z) = (x^2, y + z)$

2. Which of the following mappings of $R^2 \to R^3$ are linear? 1-1? onto?

   (a) $T(x, y) = (x + y, x - y, 3x)$
   (b) $T(x, y) = (x + y, 2x + 2y, 3x + 3y)$
   (c) $T(x, y) = (0, 0, 1)$
   (d) $T(x, y) = (0, 0, x)$
   (e) $T(x, y) = (x, -x, x)$

3. Let S, T be linear mappings of $R^2 \to R^2$, and suppose $S(1, 0) = (3, 4)$, $S(0, 1) = (5, 2)$, $T(1, 0) = (1, 1)$, $T(0, 1) = (2, 2)$. Find rules for determining the images of (x, y) under S, T, ST and TS, and classify each as 1-1, onto, neither, or both. Compare ST and TS.

4. (a) In the space P of all real polynomials, let T be the mapping that sends each polynomial $p(x)$ into $p'(x)$. Is T 1-1? onto?
   (b) In the space P let $T(p(x)) = p'(x) - 2p(x)$. Show T is linear. Is T 1-1? onto? (Hint: Find the image of the polynomial (?) $p(x) = e^{2x} \int e^{-2x} q(x) \, dx$.)

5. In the space P of all real polynomials, let $T(p(x)) = \int_0^x p(t) \, dt$. Show that T is a linear transformation of P into P. Is T 1-1? onto?

6. Let $V = C(a, b)$, the space of real-valued functions continuous on $[a, b]$, and let $T : V \to R$ by

$$T(f(x)) = \int_a^b e^{-t} f(t) \, dt$$

Linear Transformations

Show that T is linear. Is T 1-1? onto? (Hint: To show T is not 1-1, find a nonzero function $g(x)$ for which $T(g(x)) = 0$.)

7. Let T be a linear transformation of R^2 into R^2 such that $T(1, 2) = (-1, -2)$ and $T(3, 2) = (6, 4)$.
   (a) Find an expression for $T(x, y)$.
   (b) Find the matrix of T relative to the basis $\{(1, 0), (0, 1)\}$.
   (c) Find the matrix of T relative to the basis $\{(1, 2), (3, 2)\}$.
   (d) Find the matrix of T relative to the basis $\{(-1, -2), (6, 4)\}$.

8. Let T be a linear transformation of $R^n \to R^n$ with matrix A relative to the basis $\{\alpha_1, \alpha_2, \ldots, \alpha_n\}$, and suppose that $\{T(\alpha_1), \ldots, T(\alpha_n)\}$ is also a basis. Show that the matrix of T relative to the second basis is also A.

9. Let P_3 be the space of real polynomials of degree less than or equal to 3, and define $T: P_3 \to P_3$ by $T(p(x)) = p''(x) - 4p'(x) + p(x)$. Show that T is linear, and find the matrix of T relative to the basis $\{1, x, x^2, x^3\}$.

10. Let S, T be linear transformations of $R^2 \to R^2$ with $S(x, y) = (-x, 2x + y)$, $T(x, y) = (x - y, x + y)$. Find matrices representing S, T, ST, TS relative to the basis $\{(1, 0), (0, 1)\}$. Compare ST and TS and the matrices representing them.

11. Let S and T be linear transformations of R^2 into R^2 such that $S(1, 0) = (1, 1)$, $S(0, 1) = (2, 3)$, $T(1, 1) = (1, 0)$, $T(1, -1) = (5, -2)$. Find the matrix of each of the transformations S, T, ST, TS relative to the basis $\{(1, 0), (0, 1)\}$.

12. Show that the mapping $T: R^n \to R^m$ is linear if and only if T can be expressed

$$T(x_1, \ldots, x_n) = \left(\sum_{j=1}^{n} a_{1j}x_j, \sum_{j=1}^{n} a_{2j}x_j, \ldots, \sum_{j=1}^{n} a_{mj}x_j\right)$$

13. Let the general element of R^3 be the transpose of $X = [x_1 \ x_2 \ x_3]$, and let the general element of R^2 be a 2 by 1 matrix. Let $T: R^3 \to R^2$ according to the rule $T(X^T) = AX^T$, where A is a 2 by 3 matrix. Show that T is linear, and find the matrix of T relative to the standard bases.

14. Let f, g, h be mappings such that the composites $(fg)h$ and $f(gh)$ exist. Prove the associative law $(fg)h = f(gh)$ and use this to argue that matrix multiplication is associative.

15. Let $T: U \to V$, $S: V \to W$. Prove:

141

Lessons in Linear Algebra

(a) If T is 1-1 and S is 1-1, then ST is 1-1.
(b) If ST maps U onto W, then S maps V onto W.

16. Explain why two linear transformations are equal if and only if they are represented by the same matrix, relative to given bases.

Remarks on Exercises

1. (a) Linear, onto.
   (b) Onto.
   (c) Linear.
   (d) Linear.
   (e) Neither.

2. (a) Linear, 1-1.
   (b) Linear.
   (c) Neither.
   (d) Linear.
   (e) Linear.

3. Since $(x, y) = x(1, 0) + y(0, 1)$, we have $S(x, y) = xS(1, 0) + yS(0, 1) = x(3, 4) + y(5, 2)$, and $S(x, y) = (3x + 5y, 4x + 2y)$. Similarly $T(x, y) = (x + 2y, x + 2y)$. $ST(x, y) = S(x + 2y, x + 2y) = (8x + 16y, 6x + 12y)$, and $TS(x, y) = (11x + 9y, 11x + 9y)$. S is 1-1 and onto. Neither of the others is 1-1 or onto. $ST \neq TS$.

4. (a) T is not 1-1, but is onto.
   (b) To show T is 1-1, let $T(p(x)) = T(q(x))$, let $d(x) = p(x) - q(x)$, and show $d(x) = 0$. Use $d'(x) - 2d(x) = 0$ to get $d(x) = ce^{2x}$. Then show that the polynomial $d(x)$ is 0 by differentiating it repeatedly. To show T is onto, use the hint. Let $q(x)$ be any polynomial. Show that $\int e^{-2x} x^n \, dx$ is e^{-2x} times a polynomial, and then note that $p(x) = e^{2x} \int e^{-2x} q(x) \, dx$ is a polynomial. It is easy to show that $T(p(x)) = q(x)$.

5. T is 1-1, because if $T(p(x)) = T(q(x))$, then $\int_0^x p(t) \, dt = \int_0^x q(t) \, dt$, and differentiation gives $p(x) = q(x)$. T is not onto, because $T(p(x))$ is always a polynomial without a constant term. Polynomials such as $x^2 + x + 2$, $x - 5$, 4 are not in the range of T.

6. To show T is not 1-1, find a constant k such that $T(1 - ke^x) = 0$. Then $T(0) = T(1 - ke^x)$. k turns out to be $(-e^{-b} + e^{-a})/(b - a)$. T is onto. In fact, let r be any real number. Then there is a constant c such that $\int_a^b ce^{-t} \, dt = r$.

Linear Transformations

7. (a) Since $(x, y) = \frac{-2x + 3y}{4}(1, 2) + \frac{2x - y}{4}(3, 2)$, we have

$T(x, y) = \frac{-2x + 3y}{4} T(1, 2) + \frac{2x - y}{4} T(3, 2)$, or $T(x, y) = \frac{1}{4}(14x - 9y, 12x - 10y)$.

   (b) $\frac{1}{4} \begin{bmatrix} 14 & -9 \\ 12 & -10 \end{bmatrix}$.

   (c) $\begin{bmatrix} -1 & 0 \\ 0 & 2 \end{bmatrix}$.

   (d) Same as (c).

8. We are given $T(\alpha_j) = \sum_i a_{ij}\alpha_i$. Apply T to each side and use linearity.

9. $\begin{bmatrix} 1 & -4 & 2 & 0 \\ 0 & 1 & -8 & 6 \\ 0 & 0 & 1 & -12 \\ 0 & 0 & 0 & 1 \end{bmatrix}$.

10. $S: \begin{bmatrix} -1 & 0 \\ 2 & 1 \end{bmatrix} \quad T: \begin{bmatrix} 1 & -1 \\ 1 & 1 \end{bmatrix}$

    $ST: \begin{bmatrix} -1 & 1 \\ 3 & -1 \end{bmatrix} \quad TS: \begin{bmatrix} -3 & -1 \\ 1 & 1 \end{bmatrix}$

    $ST \neq TS$.

11. $S: \begin{bmatrix} 1 & 2 \\ 1 & 3 \end{bmatrix} \quad T: \begin{bmatrix} 3 & -2 \\ -1 & 1 \end{bmatrix}$

    $ST: \begin{bmatrix} 1 & 0 \\ 0 & 1 \end{bmatrix} \quad TS: \begin{bmatrix} 1 & 0 \\ 0 & 1 \end{bmatrix}$

    S and T are "inverses."

12. *Part 1.* Let $T(x_1, \ldots, x_n) = (\sum_j a_{1j}x_j, \ldots, \sum_j a_{mj}x_j)$. To show linearity, we have

$T[a(x_1, \ldots, x_n) + b(y_1, \ldots, y_n)]$
$= T(ax_1 + by_1, \ldots, ax_n + by_n)$
$= \left[\sum_j a_{1j}(ax_j + by_j), \ldots, \sum_j a_{mj}(ax_j + by_j)\right]$
$= a\left[\sum_j a_{1j}x_j, \ldots, \sum_j a_{mj}x_j\right] + b\left[\sum_j a_{1j}y_j, \ldots, \sum_j a_{mj}y_j\right]$
$= aT(x_1, \ldots, x_n) + bT(y_1, \ldots, y_n)$

(An alternate approach would be to express $T(\alpha)$ as AX^T, where $X = (x_1, \ldots, x_n)$, and then use matrix algebra to show T is linear.)

Part 2. Suppose that T is linear, and let $A = [a_{ij}]_{m,n}$ be the matrix of T relative to the standard bases $\{\alpha_i, \ldots, \alpha_n\}, \{\beta_1, \ldots, \beta_m\}$. Then

$$T(x_1, \ldots, x_n) = T\left(\sum_{j=1}^{n} x_j \alpha_j\right)$$

$$= \sum_{j=1}^{n} x_j T(\alpha_j)$$

$$= \sum_{j=1}^{n} \sum_{i=1}^{m} x_j a_{ij} \beta_i$$

$$= \sum_{i=1}^{m} \sum_{j=1}^{n} x_j a_{ij} \beta_i$$

$$= \left[\sum_{j=1}^{n} a_{1j} x_j, \sum_{j=1}^{n} a_{2j} x_j, \ldots, \sum_{j=1}^{n} a_{mj} x_j\right]$$

13. Since $T(X^T) = AX^T$, we have for all $a, b, X, Y, T(aX^T + bY^T) = A(aX^T + bY^T) = aAX^T + bAY^T = aT(X^T) + bT(Y^T)$, and T is linear. The matrix of T relative to the standard bases is A.

14. Let $h : U \to V, g : V \to W, f : W \to X$. Using repeatedly the definition of composition of functions, we have for all $u \in U$, $[(fg)h](u) = (fg)[h(u)] = f[g(h(u))] = f[(gh)(u)] = [f(gh)](u)$. Therefore, $(fg)h = f(gh)$. Since the multiplication of two matrices represents the composition of two linear transformations, it follows that matrix multiplication is associative.

15. (a) Let $u_1, u_2 \in U$. Then

    $u_1 \neq u_2 \Rightarrow T(u_1) \neq T(u_2)$ (T is 1-1)

    $\Rightarrow ST(u_1) \neq ST(u_2)$ (S is 1-1)

    $\Rightarrow ST$ is 1-1

    (b) Let $w \in W$. Then there is an element $u \in U$ such that $ST(u) = w$, because ST maps U onto W. This means there is an element in V, namely $T(u)$, whose image under S is w. Therefore, S maps V onto W.

13

RANK, NULLITY, AND INVERSE OF A LINEAR TRANSFORMATION

For a linear transformation T we introduce and discuss the concepts of *image space*, *kernel*, *rank*, and *nullity*, and we interpret these in relation to T and to any matrix representing T. The results have bearing on a wide variety of "linear problems" arising from differential equations, systems of equations, numerical analysis, physics, computing science, and others.

We conclude the lesson by showing that every finite-dimensional linear space has the same form as (is isomorphic to) an n-tuple space.

A linear transformation $T: V \to W$ has the property that it links the subspaces of V with those of W.

13-1 Theorem

Let T be a linear transformation of V into W.

(a) If V_1 is a subspace of V, then $T(V_1)$, the set of all images of vectors in V_1, is a subspace of W.

(b) If W_1 is a subspace of W, then $T^{-1}(W_1)$, the set of all inverse images of vectors in W_1, is a subspace of V.

Proof: (a) Let β_1, β_2 be typical elements of $T(V_1)$. Then $\beta_1 = T(\alpha_1)$, $\beta_2 = T(\alpha_2)$ where $\alpha_1, \alpha_2 \in V_1$. To show that $T(V_1)$ is a subspace of W we show that $T(V_1)$ is closed by noting that

$$a_1\beta_1 + a_2\beta_2 = a_1 T(\alpha_1) + a_2 T(\alpha_2)$$
$$= T(a_1\alpha_1 + a_2\alpha_2) \quad \text{(linearity of } T\text{)}$$
$$\in T(V_1) \quad (V_1 \text{ is a subspace.})$$

(b) Now let α_1, α_2 be any two vectors in $T^{-1}(W_1)$, that is, let $T(\alpha_1) = \beta_1$, $T(\alpha_2) = \beta_2$ where $\beta_1, \beta_2 \in W_1$. We show $T^{-1}(W_1)$ is a subspace of V by showing that $T^{-1}(W_1)$ is closed.

Lessons in Linear Algebra

$$T(a_1\alpha_1 + a_2\alpha_2) = a_1T(\alpha_1) + a_2T(\alpha_2)$$
$$= a_1\beta_1 + a_2\beta_2 \; \varepsilon \; W_1$$

Therefore, $a_1\alpha_1 + a_2\alpha_2 \; \varepsilon \; T^{-1}(W_1)$. ∎

We summarize by remarking that under T, subspaces of V correspond to subspaces of W, but not necessarily in 1-1 fashion. A special case is that $T(0) = 0$.

Example 1. Let $T : R^3 \to R^2$ according to $T(x, y, z) = (x - y, x + y)$, and let $V_1 = \{(x, x, x)\}$ and $W_1 = \{(x, -x)\}$ be subspaces of R^3 and R^2.

Since $T(x, x, x) = (0, 2x)$, we have $T(V_1)$ is the space of multiples of $(0, 1)$.

Also, the set of inverse images of $(x, -x)$ is $\{(0, -x, z) \mid z \in R\}$, a subspace of R^3.

As special cases of Theorem 13-1, we have $T(V)$ and $T^{-1}(0)$ are subspaces of W and V, respectively. These two subspaces are given special names.

13-2 Definition

Let T be a linear transformation of V into W. Then

(a) the *image space of T* is $T(V)$.
(b) the *kernel of T* (or *null space of T*) is $T^{-1}(0)$.
(c) the *rank of T* is the dimension of the image space.
(d) the *nullity of T* is the dimension of the kernel.

In Example 1 we have $T(x, y, z) = (x - y, x + y)$. So the image space of T is all of R^2, and the kernel of T is the set $\{(0, 0, z)\}$.

Example 2. Let $T : R^3 \to R^3$ by $T(x, y, z) = (x - y, y - z, z - x)$. A vector (u, v, w) is in $T(R^3)$ if and only if $(u, v, w) = (x - y, y - z, z - x)$ for some x, y, z. This leads to the conclusion that $T(V) = \{(u, v, w) \mid u + v + w = 0\}$. A basis of $T(V)$ is $\{(1, 0, -1), (-1, 1, 0)\}$, and the rank of T is 2.

The kernel of T is $\{(x, y, z) \mid x - y = y - z = z - x = 0\}$ or the set $\{(x, x, x)\}$. A basis for the kernel is $\{(1, 1, 1)\}$, and the nullity of T is 1.

Observe that the rank and nullity add up to 3, a fact which we shall generalize in Theorem 13-3.

Rank, Nullity, and Inverse of a Linear Transformation

Example 3. Let $T: P \to P$ where $T(p(x)) = p'(x)$. (As usual, P is the space of real polynomials.)

Then the image space $T(P) = P$ because any polynomial $q(x)$ is the image of some $p(x)$. The kernel of T is the set of constant polynomials. The rank of T is infinite, the nullity of T is 1.

13-3 Theorem

Let T be a linear transformation of V into W, and let dim $V = n$. Then rank (T) + nullity $(T) = n$.

Proof: Let ρ = rank (T), ν = nullity (T), let $\{\alpha_1, \ldots, \alpha_\nu\}$ be a basis of the kernel, and let $\{\alpha_1, \ldots, \alpha_\nu, \ldots, \alpha_n\}$ be a basis of V. It is easy to show that $T(V)$ is spanned by the set $\{T(\alpha_{\nu+1}), \ldots, T(\alpha_n)\}$ (see Exercise 9). Also the set is linearly independent, because if $\sum_{i=\nu+1}^{n} c_i T(\alpha_i) = 0$ nontrivially we would have $T(\sum_{i=\nu+1}^{n} c_i \alpha_i) = 0$ contradicting the assumption that $\{\alpha_1, \ldots, \alpha_\nu\}$ is a basis of the kernel.

Therefore, $\rho = n - \nu$, and $\rho + \nu = n$. ∎

Examples 1 and 2 can be used to illustrate Theorem 13-3, for in each case $\rho(T) + \nu(T) = n$ because $2 + 1 = 3$. Theorem 13-3 does not apply to Example 3 because dim $P = \infty$. However, Theorem 13-3 requires only that dim V be finite. No such requirement is placed on W.

Example 4. Let V be the function space spanned by the basis $\{1, x, x^2, \sin x, \cos x\}$, and let W be the space of all real functions. Define $D: V \to W$ by $T(f) = D^2 f$, the second derivative of f.

Then $T(V)$ is the set spanned by $\{1, \sin x, \cos x\}$ and $\rho(T) = 3$. Furthermore, the kernel of T is spanned by $\{1, x\}$, and $\nu(T) = 2$. Note that $3 + 2 = 5$, the dimension of V.

We now return to the representation of a linear transformation T by a matrix A. We will show, among other things, that the rank of T equals the rank of A.

13-4 Theorem

Let V, W be two linear spaces with bases $\mathcal{A} = \{\alpha_1, \ldots, \alpha_n\}$ and $\mathcal{B} = \{\beta_1, \ldots, \beta_m\}$, let T be a linear transformation of V into W, and let $A = [a_{ij}]_{m,n}$ be the matrix of T relative to the bases $\mathcal{A}, \mathcal{B}$. Let $\alpha = \dot{x}_1 \alpha_1 + \ldots + x_n \alpha_n$ and $\beta = y_1 \beta_1 + \ldots + y_m \beta_m$. Then

Lessons in Linear Algebra

(a) $T(\alpha) = \beta$ if and only if $AX^T = Y^T$, where $X = [x_1 \ldots x_n]$, $Y = [y_1 \ldots y_m]$.
(b) rank (T) = rank (A).

Proof: (a)

$$T(\alpha) = T\left(\sum_{j=1}^{n} x_j \alpha_j\right) \quad \text{(given)}$$

$$= \sum_{j=1}^{n} x_j T(\alpha_j) \quad \text{(linearity)}$$

$$= \sum_{j=1}^{n} x_j \sum_{i=1}^{m} a_{ij} \beta_i \quad \text{(definition of matrix of } T\text{)}$$

$$= \sum_{i=1}^{m} \sum_{j=1}^{n} a_{ij} x_j \beta_i \quad \text{(properties of summations)}$$

Also,

$$\beta = \sum_{i=1}^{m} y_i \beta_i \quad \text{(given)}$$

Therefore, $T(\alpha) = \beta$ if and only if $\sum_{j=1}^{n} a_{ij} x_j = y_i$ for $i = 1, \ldots, m$. This is equivalent to the matrix equation $AX^T = Y^T$.

(b) The rank of T is the dimension of the space spanned by $T(\alpha_1), \ldots, T(\alpha_n)$. Each one of these vectors is a linear combination of $\{\beta_1, \ldots, \beta_m\}$, specifically $T(\alpha_j) = \sum_{i=1}^{m} a_{ij} \beta_i$. Each linear relation of $\{T(\alpha_1), \ldots, T(\alpha_n)\}$ corresponds to a linear relation of the columns of A, and the rank of T is therefore the rank of A. ∎

Example 5. Let $T: R^2 \to R^2$ with $T(x, y) = (x + y, x + y)$, and choose some vector, for example $\alpha = (1, 3)$. Then $T(\alpha) = (4, 4)$.

Relative to the basis $\{(1, 0), (0, 1)\}$:

T is represented by $A = \begin{bmatrix} 1 & 1 \\ 1 & 1 \end{bmatrix}$.

α is represented by $X = [1 \quad 3]$.

$T(\alpha)$ is represented by $Y = [4 \quad 4]$, $Y^T = AX^T$.

Relative to the basis $\{(1, 1), (1, -1)\}$:

Rank, Nullity, and Inverse of a Linear Transformation

T is represented by $B = \begin{bmatrix} 2 & 0 \\ 0 & 0 \end{bmatrix}$.

α is represented by $X = [2 \ -1]$.

$T(\alpha)$ is represented by $Y = [4 \ 0]$, $Y^T = BX^T$.

We say $(x_1, \ldots, x_n)$ are the "coordinates" of α relative to the basis $\{\alpha_1, \ldots, \alpha_n\}$ provided that $\alpha = x_1\alpha_1 + \ldots + x_n\alpha_n$.

If T is a 1-1 mapping of a set V onto a set W, then there is a 1-1 inverse mapping T^{-1} of W onto V. Given such a T, define T^{-1} by $T^{-1}(\beta) = \alpha$ if and only if $T(\alpha) = \beta$. Then T^{-1} will be a 1-1 mapping of W onto V. Furthermore, $TT^{-1} = 1_w$, the identity mapping on W, and $T^{-1}T = 1_v$ (Exercise 12).

In case V and W are linear spaces, there are other conclusions.

13-5 Theorem

Let V and W be linear spaces with bases $\mathcal{C} = \{\alpha_1, \ldots, \alpha_n\}$ and $\mathcal{B} = \{\beta_1, \ldots, \beta_n\}$, respectively, and let T be an invertible linear transformation of V onto W. Then

(a) T^{-1} is linear.
(b) if A is the matrix of T relative to the bases $\mathcal{C}$, $\mathcal{B}$, then A^{-1} is the matrix of T^{-1} relative to the bases $\mathcal{B}$, $\mathcal{C}$.

Proof: (a) We are given that T is linear and that T^{-1} exists. To show T^{-1} is linear, let β_1, β_2 be any two vectors in W, with $T(\alpha_1) = \beta_1$, and $T(\alpha_2) = \beta_2$. Then

$$\begin{aligned} T^{-1}(a_1\beta_1 + a_2\beta_2) &= T^{-1}[a_1T(\alpha_1) + a_2T(\alpha_2)] \\ &= T^{-1}T(a_1\alpha_1 + a_2\alpha_2) \quad (T \text{ is linear}) \\ &= 1_v(a_1\alpha_1 + a_2\alpha_2) \quad (T^{-1}T = 1_v) \\ &= a_1\alpha_1 + a_2\alpha_2 \\ &= a_1T^{-1}(\beta_1) + a_2T^{-1}(\beta_2) \end{aligned}$$

(b) If A is the matrix of T, then A^{-1} must be the matrix of T^{-1} so that $AA^{-1} = A^{-1}A = I_n$. ∎

Example 6. Let $T: R^2 \to R^2$ by $T(x, y) = (x, x + y)$. Then $T^{-1}(x, y) = (x, y - x)$, and $TT^{-1} = T^{-1}T = 1$. Relative to the standard bases the matrices representing T and T^{-1} are the inverse matrices

$$\begin{bmatrix} 1 & 0 \\ 1 & 1 \end{bmatrix} \text{ and } \begin{bmatrix} 1 & 0 \\ -1 & 1 \end{bmatrix}$$

Lessons in Linear Algebra

13-6 Theorem Let V and W be n-dimensional linear spaces, and let $T: V \to W$ be linear. The following are equivalent:
  (a) T is invertible.
  (b) T is onto.
  (c) rank $(T) = n$.
  (d) nullity $(T) = 0$.
  (e) kernel $(T) = \{0\}$.
  (f) T is 1-1.

Proof: The plan is to show (a) $\Rightarrow$ (b) $\Rightarrow$ (c) $\Rightarrow$ (d) $\Rightarrow$ (e) $\Rightarrow$ (f) $\Rightarrow$ (a).
(a) $\Rightarrow$ (b), because if T were not a mapping of V onto W, then we could not find T^{-1} satisfying $TT^{-1} = 1_w$. (b) $\Rightarrow$ (c), because if T is onto, then $T(V) = W$, and rank $T = \dim W = n$. (c) $\Rightarrow$ (d), because rank + nullity $= n$, by Theorem 13-3. (d) $\Rightarrow$ (e), since the nullity is by definition the dimension of the kernel. (e) $\Rightarrow$ (f), for suppose kernel $T = \{0\}$. Then $T(\alpha_1) = T(\alpha_2) \Rightarrow T(\alpha_1 - \alpha_2) = 0 \Rightarrow \alpha_1 - \alpha_2 \,\varepsilon\, \text{kernel}\, T \Rightarrow \alpha_1 = \alpha_2$, and T is 1-1. (f) $\Rightarrow$ (a), for suppose T is 1-1. Then the kernel of T is $\{0\}$ and the rank of T is n, by Theorem 13-3. Therefore, T is onto and hence invertible. ∎

In practice we use any one of the six conditions to test T for invertibility.

Example 7. Let $T: R^3 \to R^3$ according to $T(x, y, z) = (x + y, y + z, x - z)$. To test for invertibility, we notice that the kernel contains the nonzero vector $(1, -1, 1)$. We decide immediately that T is not invertible. We may also note that kernel of $T = \{(t, -t, t)\}$, nullity $T = 1$, rank $T = 2$, T is not onto, and T is not 1-1. The point is that any one of these observations shows that T is not invertible.

Example 8. Determine whether or not the matrix

$$A = \begin{bmatrix} 1 & 2 & 3 \\ -1 & 2 & 4 \\ 1 & 6 & 10 \end{bmatrix}$$

is invertible.
Let $X = [2 \ -7 \ 4]$ and note that $AX^T = 0$. Then A represents a linear transformation with a nonzero kernel and, therefore, A is not invertible.

Rank, Nullity, and Inverse of a Linear Transformation

13-7 Definition

Let V and W be two linear spaces over the same field. V is *isomorphic* to W if there is a 1-1 linear transformation of V onto W.

Isomorphism is an equivalence relation because it is reflexive, symmetric, and transitive. The identity mapping shows that V is isomorphic to itself. Next, if T is an isomorphism of V onto W, then T^{-1} is an isomorphism of W onto V. Finally, if $T: V \to W$ and $S: W \to X$ are isomorphisms, then so is $ST: V \to X$.

The linear space we have used most frequently is the n-tuple space. The study of such spaces is equivalent to the study of all finite-dimensional spaces.

13-8 Theorem

Let V be an n-dimensional linear space over F. Then V is isomorphic to F^n, the space of n-tuples of elements in F.

Proof: Let $\{\alpha_1, \ldots, \alpha_n\}$ be a basis of V. The general vector α of V can be written uniquely as $\alpha = x_1\alpha_1 + \ldots + x_n\alpha_n$. Define $T: V \to F^n$ by

$$T(\alpha) = (x_1, \ldots, x_n)$$

Then T is a 1-1 mapping of V onto F^n. Also, T is linear because if $\beta = y_1\alpha_1 + \ldots + y_n\alpha_n$ is another vector, then

$$\begin{aligned} T(a\alpha + b\beta) &= T\Big[a \sum x_i\alpha_i + b \sum y_i\alpha_i\Big] \\ &= T\Big[\sum (ax_i + by_i)\alpha_i\Big] \\ &= (ax_1 + by_1, \ldots, ax_n + by_n) \quad \text{(definition of } T\text{)} \\ &= a(x_1, \ldots, x_n) + b(y_1, \ldots, y_n) \\ &= aT(\alpha) + bT(\beta) \quad \blacksquare \end{aligned}$$

Summary of Topics in Lesson 13

The image of a subspace is a subspace, and the inverse image of a subspace is a subspace.

Image space, kernel, rank, and nullity of a linear transformation.

Representing the linear transformation T by $T(\alpha) = \beta$ or by the matrix equation $AX^T = Y^T$.

The rank of T equals the rank of any matrix representing T.

The inverse of a linear transformation. Conditions for invertibility.

Any finite-dimensional linear space is isomorphic to a space of n-tuples.

Lessons in Linear Algebra

Exercises

1. For each of the following linear transformations T find the image space of T, the kernel of T, the rank of T, and the nullity of T.
   (a) $T: R^3 \to R^2$, $T(x, y, z) = (x - 2y, 2x + z)$.
   (b) $T: R^3 \to R^3$, $T(x, y, z) = (x, x + y + z, y + z)$.
   (c) $T: R^4 \to R^3$, $T(x, y, z, w) = (y, 2y, 3y)$.
   (d) $T: P \to P$, $T(p) = p''$.
   (e) Let V be the space of real functions continuous on $(-\infty, \infty)$ and let W be the space of all real functions. Let $T: V \to W$,
   $$T(f(x)) = \int_a^x f(t)\, dt.$$

2. Let $T: R^2 \to R^3$, $T(x, y) = (x + y, x - y, x + y)$. For $\alpha = (3, 5)$, we have $T(\alpha) = (8, -2, 8)$.
   (a) Find the matrix of T relative to the standard bases and check that $T(\alpha) = \beta$ is represented $AX^T = Y^T$.
   (b) Repeat (a), using the bases $\{(1, 1), (1, -1)\}$ and $\{(0, 1, 1), (1, 0, 1), (1, 1, 0)\}$.
   (c) Find the rank and nullity of T.

3. Let T be a linear transformation of V into W.
   (a) Show, without appealing to Theorem 13-1, that $T(0) = 0$.
   (b) Show that if two distinct elements of V have the same image, then their difference is in the kernel of T.
   (c) Suppose $T(\alpha_1) = T(\alpha_2) = T(\alpha_3) = \beta$. How would you choose c_1, c_2, c_3 in order that $T(c_1\alpha_1 + c_2\alpha_2 + c_3\alpha_3) = \beta$?

4. (a) Let $T: R^2 \to R^2$, $T(x, y) = (x - y, 2x - 2y)$ and let V be the subspace $\{(t, 3t)\}$. Find $T(V)$ and $T^{-1}(V)$. Also find another subspace W such that $T(V) = T(W)$. Observe that T is not invertible.
   (b) Let $T: R^2 \to R^2$, $T(x, y) = (x - y, x - 2y)$, and let V be the subspace $\{(t, t)\}$. Find $T(V)$ and $T^{-1}(V)$ by finding the rule for $T^{-1}(x, y)$.

5. If T is a linear transformation with $T(\alpha) = \beta$ and $T(\alpha_1) = 0$, then $T(\alpha + \alpha_1) = \beta$. Interpret this for each of the following linear transformations:
   (a) Since $(2, 1)$ is a solution of $2x + y = 5$, and $(1, -2)$ is a solution of $2x + y = 0$, then $(3, -1)$ is a solution of what equation?
   (b) Since x is a solution of the differential equation $y'' - 3y' + 2y = 2x - 3$, and e^x is a solution of $y'' - 3y' + 2y = 0$, find another solution of the first differential equation.

Rank, Nullity, and Inverse of a Linear Transformation

6. Let $T: P_2 \to R^3$ according to the rule $T(ax^2 + bx + c) = (a + b + c, b + c, c)$.
   (a) Find the matrix A of T relative to the bases $\{1, x, x^2\}$ and $\{(1, 0, 0), (0, 1, 0), (0, 0, 1)\}$.
   (b) Find A^{-1} and the rule for $T^{-1}(a, b, c)$.

7. Let $T: R^3 \to R^3$, $T(x, y, z) = (x, y + z, x + y + z)$.
   (a) Find the matrix A of T relative to the standard basis, and determine whether or not A and T are invertible.
   (b) Find the rank and nullity of A.

8. (a) Prove that if A is an invertible matrix and n is a positive integer, then $(A^{-1})^n = (A^n)^{-1}$.
   (b) Let T be an invertible linear transformation of V into V. Explain why $(T^n)^{-1} = (T^{-1})^n$ for n a positive integer. Here T^n means $TT \ldots T$, a composition.
   (c) Given $T: R^2 \to R^2$, $T(x, y) = (x, x + y)$, find the rules for T^2, T^{-1}, $(T^2)^{-1}$, $(T^{-1})^2$.

9. In the proof of Theorem 13-3, the transformation $T: V \to W$ had nullity ν. A basis of V is $\{\alpha_1, \ldots, \alpha_\nu \ldots, \alpha_n\}$, where $\{\alpha_1, \ldots, \alpha_\nu\}$ is a basis of the kernel of T. Show that the set $T(\alpha_{\nu+1}), \ldots, T(\alpha_n)$ does span $T(V)$, as was stated in the proof of the theorem.

10. Consider the linear transformations $T: U \to V$, $S: V \to W$, where U, V, and W are finite-dimensional.
    (a) Prove that kernel $T \subset$ kernel ST, and hence, $\nu(T) \leq \nu(ST)$.
    (b) Compare the image spaces of ST and T, and show that $\rho(ST) \leq \rho(T)$.
    (c) Show $\rho(ST) \leq \rho(S)$.
    (d) Use the above to argue that the rank of the product of two matrices cannot exceed the rank of either matrix.

11. For the transformations in Exercise 10:
    (a) Prove that $\rho(T) \leq \rho(ST) + \nu(S)$. (Hint: Let S' be the transformation S, restricted to $T(U)$. Use Theorem 13-3 to show that $\rho(S') + \nu(S') = \rho(T)$.)
    (b) Show that $\nu(ST) \leq \nu(T) + \nu(S)$.

12. Let T be a 1-1 mapping of V onto W, and show that $T^{-1}T = 1_v$, the identity mapping on V and that $TT^{-1} = 1_w$.

13. Explain why the subspace $(x, y, z - w, x - z + w)$ of R^4 is isomorphic to R^3.

153

Lessons in Linear Algebra

14. Let T be a linear transformation of the infinite-dimensional linear space V into W. Prove that at least one of the pair $\rho(T)$, $\nu(T)$ must be infinite. (Assume both ρ, ν finite and look at the set $\{T(\alpha_1), \ldots, T(\alpha_{\nu+\rho+1})\}$, where $\{\alpha_1, \ldots, \alpha_\nu\}$ is a basis for the kernel of T and $\{\alpha_1, \ldots, \alpha_{\nu+\rho+1}\}$ is linearly independent.)

Remarks on Exercises

1. (a) Image space $= R^2$. kernel $= \{$multiples of $(2, 1, -4)\}$. rank $= 2$. nullity $= 1$.
   (b) The image space has basis $\{(1, 1, 0), (0, 1, 1)\}$, so the rank of T is 2. The kernel has basis $\{(0, 1, -1)\}$ and the nullity of T is 1.
   (c) The image space has basis $\{(1, 2, 3)\}$, so the rank of T is 1. The kernel has basis $\{(1, 0, 0, 0), (0, 0, 1, 0), (0, 0, 0, 1)\}$, and the nullity of T is 3.
   (d) The image space is P. The kernel is P_1, the set of polynomials of degree less than or equal to 1. Therefore, the nullity of T is 2, and the rank is infinite.
   (e) The image space is the set of all functions F such that $F'(x)$ is continuous and $F(a) = 0$, so the rank of T is infinite. The kernel is the zero function, so the nullity of T is 0. Use the fact that $F'(x) = f(x)$.

2. (a) $A = \begin{bmatrix} 1 & 1 \\ 1 & -1 \\ 1 & 1 \end{bmatrix}$, $X^T = \begin{bmatrix} 3 \\ 5 \end{bmatrix}$, $Y^T = \begin{bmatrix} 8 \\ -2 \\ 8 \end{bmatrix}$.

   (b) $A = \begin{bmatrix} 0 & 1 \\ 2 & -1 \\ 0 & 1 \end{bmatrix}$, $X^T = \begin{bmatrix} 4 \\ -1 \end{bmatrix}$, $Y^T = \begin{bmatrix} -1 \\ 9 \\ -1 \end{bmatrix}$.

   (c) $\rho(T) = 2$, $\nu(T) = 0$.

3. (a) Consider $T(\alpha + 0)$.
   (b) Let $T(\alpha_1) = T(\alpha_2) = \beta$, and look at $T(\alpha_1 - \alpha_2)$.
   (c) $c_1 + c_2 + c_3 = 1$.

4. (a) $T(V) = \{t(1, 2)\}$, the space spanned by $(1, 2)$. $T^{-1}(V) = \{t(1, 1)\}$. W can be any subspace other than the kernel.
   (b) $T(V)$ is the space spanned by $(0, 1)$. $T^{-1}(x, y) = (2x - y, x - y)$. $T^{-1}(V)$ is the space spanned by $(1, 0)$.

5. (a) $2x + y = 5$.
   (b) $x + e^x$.

6.
(a) $A = \begin{bmatrix} 1 & 1 & 1 \\ 1 & 1 & 0 \\ 1 & 0 & 0 \end{bmatrix}$.

(b) $A^{-1} = \begin{bmatrix} 0 & 0 & 1 \\ 0 & 1 & -1 \\ 1 & -1 & 0 \end{bmatrix}$. $T^{-1}(a, b, c) = (a - b)x^2 + (b - c)x + c$.

7. (a) $A = \begin{bmatrix} 1 & 0 & 0 \\ 0 & 1 & 1 \\ 1 & 1 & 1 \end{bmatrix}$, a noninvertible matrix.

(b) $\rho(A) = 2$, $\nu(A) = 1$.

8. (a) Use mathematical induction.
(b) Refer to part (a).
(c) $T^2(x, y) = (x, 2x + y)$, $T^{-1}(x, y) = (x, -x + y)$, $(T^{-1})^2(x, y) = (T^2)^{-1}(x, y) = (x, -2x + y)$.

9. The general element of $T(V)$ is

$$T\left(\sum_{i=1}^{n} a_i \alpha_i\right) = T\left(\sum_{i=1}^{\nu} a_i \alpha_i\right) + T\left(\sum_{i=\nu+1}^{n} a_i \alpha_i\right)$$

$$= T\left(\sum_{i=\nu+1}^{n} a_i \alpha_i\right)$$

10. Venn diagrams are helpful.
(a) $\alpha \varepsilon$ kernel $T \Rightarrow T(\alpha) = 0 \Rightarrow ST(\alpha) = 0 \Rightarrow \alpha \varepsilon$ kernel (ST). Since kernel $T \subset$ kernel (ST), we have $\nu(T) \leq \nu(ST)$.
(b) Consider $S_{T(U)}$, that is the transformation S restricted to $T(U)$. Applying Theorem 13-3 gives $\rho(ST) + \nu(S_{T(U)}) = \dim T(U)$, and $\rho(ST) \leq \rho(T)$.
(c) If $\gamma = (ST)(\alpha)$, then $\gamma = S(T(\alpha))$. That is, the image space of ST is a subset of the image space of S, and therefore, $\rho(ST) \leq \rho(S)$.
(d) Let A, B be two matrices, such that AB exists. Then A, B, AB represent linear transformations S, T, ST (relative to some bases), and

$$\rho(AB) = \rho(ST) \leq \min \{\rho(S), \rho(T)\} = \min \{\rho(A), \rho(B)\}$$

11. (a) Since $\rho(S') + \nu(S') = \rho(T)$, and $\nu(S') \leq \nu(S)$, we have $\rho(ST) + \nu(S) \geq \rho(T)$.
(b) Let $\dim U = n$. By part (a) and Theorem 13-3, $n - \nu(ST) + \nu(S) \geq n - \nu(T)$, etc.

12. Let $\alpha \in V$, with $T(\alpha) = \beta$. Then $T^{-1}T(\alpha) = T^{-1}(\beta) = \alpha$, and $T^{-1}T = 1_v$. Now, let $\beta \in W$, with $T(\alpha) = \beta$. Then $TT^{-1}(\beta) = T(\alpha) = \beta$ and $TT^{-1} = 1_w$.

13. The subspace can be given as $\{x(1, 0, 0, 1) + y(0, 1, 0, 0) + z(0, 0, 1, -1) + w(0, 0, -1, 1)\}$. A basis for the subspace is $\{(1, 0, 0, 1), (0, 1, 0, 0), (0, 0, 1, -1)\}$, and the dimension is 3. By Theorem 13-8, the subspace is isomorphic to R^3.

14

THE LINEAR PROBLEM—
MORE OPERATIONS ON
LINEAR TRANSFORMATIONS

There are two main topics in this lesson. First, we describe the "linear problem." For it we list and prove a number of general results, which are immediately applied to four different problems. You should observe that the general proofs are quite simple, even though they apply forcefully to many types of linear problems. You should enjoy anticipating and proving these results.

Second, we define addition and scalar multiplication of linear transformations and we note that the algebra of linear transformations of finite-dimensional linear spaces is equivalent to the algebra of m by n matrices. Of course, we have already done this for the composition (multiplication) of linear transformations.

Let $T: V \to W$ be linear, and let β be some vector in W (see Figure 14-1).

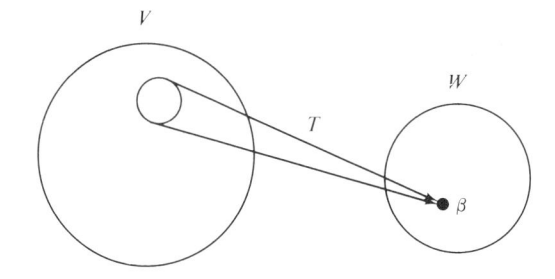

14-1
Definition

FIGURE 14-1

The *linear problem* is that of finding all $\alpha \in V$ such that $T(\alpha) = \beta$. If $\beta = 0$, the problem is called *homogeneous*; if $\beta \neq 0$, the problem is called *nonhomogeneous*.

Lessons in Linear Algebra

It is clear that the solution set of the homogeneous problem $T(\alpha) = 0$ is the kernel of T. The following results are helpful in solving any linear problem.

14-2 Theorem

Let $T : V \to W$ be linear. Let P represent the linear problem $T(\alpha) = \beta$ and let H represent the corresponding homogeneous problem $T(\alpha) = 0$. Then

(a) the solution set of H is a subspace of V.
(b) any solution of P added to a solution of H gives a solution of P.
(c) the difference of any two solutions of P is a solution of H.
(d) let α_1 be a particular solution of P. Then every solution of P is given by $\alpha_1 + \alpha$, where α is some solution of H. That is, the solution set of P is $\alpha_1 + $ kernel (T).
(e) let $\{\alpha_1, \ldots, \alpha_r\}$ be solutions of the nonhomogeneous problem $T(\alpha) = \beta$. Then $\sum c_i \alpha_i$ is also a solution if and only if $\sum c_i = 1$.

Proof: (a) We note that the solution set of H is the kernel of T, a subspace of V.

(b) Let α_1, α_2 be solutions of P and H, that is, let $T(\alpha_1) = \beta$, $T(\alpha_2) = 0$. By linearity $T(\alpha_1 + \alpha_2) = T(\alpha_1) + T(\alpha_2) = \beta + 0 = \beta$.

(c) If $T(\alpha_1) = T(\alpha_2) = \beta$, then $T(\alpha_1 - \alpha_2) = 0$.

(d) We are given $T(\alpha_1) = \beta$, and by part (b), if $T(\alpha) = 0$, then $T(\alpha + \alpha_1) = \beta$. What we now need to show is that *any* solution α_2 of P can be written as $\alpha_1 + \alpha$, where $T(\alpha) = 0$. To this end, note that $\alpha_2 = \alpha_1 + (\alpha_2 - \alpha_1)$, where $T(\alpha_2 - \alpha_1) = 0$ [part (c)], and we are finished.

(e) This is left for you. ∎

Next we illustrate all parts of Theorem 14-2 by means of four examples.

Example 1. This example is easy to solve without Theorem 14-2 but nevertheless illustrates the theorem well. Solving the equation $2x + 3y = 6$ is a linear problem P. The corresponding homogeneous problem is $2x + 3y = 0$ (see Figure 14-2). Note that the problems can be written

$$[2 \ 3]\begin{bmatrix} x \\ y \end{bmatrix} = 6 \quad \text{and} \quad [2 \ 3]\begin{bmatrix} x \\ y \end{bmatrix} = 0$$

(a) The solution set of H is the subspace $\{t(3, -2) \mid t \in R\}$, or the points on the line $2x + 3y = 0$. (b) One solution of P is $\alpha_1 = (2, 2/3)$

The Linear Problem — More Operations on Linear Transformations

and a solution of H is $\alpha_2 = (3, -2)$. Then $\alpha_1 + \alpha_2 = (5, -4/3)$, a solution of P. (c) Let $\alpha_1 = (-3, 4)$ and $\alpha_2 = (0, 2)$, each a solution of P. Then $\alpha_1 - \alpha_2 = (-3, 2)$, a solution of H. (d) The general solution of P is $(x, (6 - 2x)/3)$, or $(0, 2) + x(1, -2/3)$, where $(0, 2)$ is a particular solution of P and $x(1, -2/3)$ is the general solution of H.

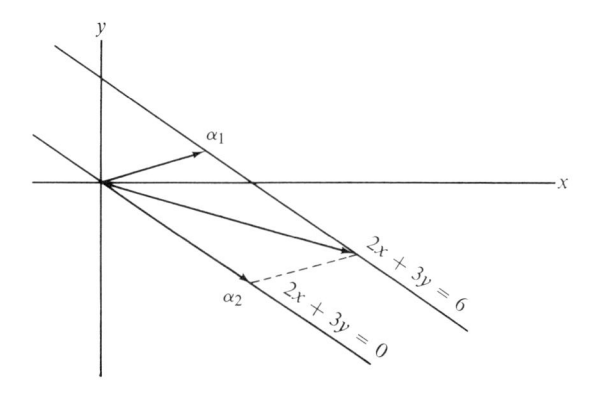

FIGURE 14-2

Also the general solution of P can be written $(3, 0) + y(-3/2, 1)$, or as $(-3, 4) + t(3, -2)$, or in many other ways. In each case the general solution of P is the sum of a particular solution of P and the general solution of H.

(e) Consider $\alpha_1 = (3, 0)$, $\alpha_2 = (0, 2)$, $\alpha_3 = (-3, 4)$, three solutions of P. Choose three scalars $c_1 = 2$, $c_2 = -5$, $c_3 = 4$ whose sum is 1. Then $\sum c_i \alpha_i = (-6, 6)$, a solution of P.

Example 2. The results of Theorem 14-2 apply also to systems of linear equations. The system

$$x + 2y - z = 2$$
$$2x + 3y + z = 11$$

can be recognized as a linear problem

$$\begin{bmatrix} 1 & 2 & -1 \\ 2 & 3 & 1 \end{bmatrix} \begin{bmatrix} x \\ y \\ z \end{bmatrix} = \begin{bmatrix} 2 \\ 11 \end{bmatrix}$$

or

$$T(\alpha) = \beta$$

where $T: R^3 \to R^2$ by means of the matrix
$$A = \begin{bmatrix} 1 & 2 & -1 \\ 2 & 3 & 1 \end{bmatrix}$$
Our problem is to find
$$\alpha = \begin{bmatrix} x \\ y \\ z \end{bmatrix}$$
such that $T(\alpha) = \beta$, where
$$\beta = \begin{bmatrix} 2 \\ 11 \end{bmatrix}$$

The rank of A is 2, so the nullity of T is 1. Hence, the vector $(5, -3, -1)$ is a basis for the kernel of T, which means the solution space of H is $\{t(5, -3, -1) \mid t \varepsilon R\}$. A particular solution of P is $(16, -7, 0)$, obtained by setting $z = 0$ and solving for x and y. Hence, the general solution of P is $\{(16, -7, 0) + t(5, -3, -1) \mid t \varepsilon R\}$. The solution could be written in other ways.

Example 3. Solving the differential equation $y' + y = 2x$ is a linear problem $T(y) = 2x$, where $T(y) = y' + y$. A particular solution of P is $2x - 2$, and the general solution of H is ce^{-x}. Hence, the general solution of the differential equation $y' + y = 2x$ is $y = 2x - 2 + ce^{-x}$.

Theorem 14-2 applies also to the nth order linear differential equation $y^{(n)} + a_1 y^{(n-1)} + \ldots + a_n y = f(x)$.

Example 4. Find all functions of the form $ax + b$ for which $\int_0^1 f(x)\,dx = 1$. Here we consider $T: P_1 \to R$ with $T(f) = \int_0^1 f(x)\,dx$. Clearly, T is linear. (A linear mapping of a linear space into its field of scalars is called a *linear functional.*)

Our problem P is to find a, b such that $T(ax + b) = 1$. Integration gives $(a/2) + b = 1$ or $a = 2 - 2b$. The general solution of P is therefore
$$(2 - 2b)x + b$$
where b is any real number.

A particular solution of P is the function 1, since $\int_0^1 1\,dx = 1$. Notice that
$$(2 - 2b)x + b = [1] + [(2 - 2b)x + b - 1]$$

The Linear Problem — More Operations on Linear Transformations

where 1 is a solution of P and $(2 - 2b)x + b - 1$ is the general solution $2tx - t$ of H, with $t = 1 - b$. (See Figure 14-3.)

Now we proceed to the second part of this lesson and define new operations, addition and scalar multiplication, for linear transformations.

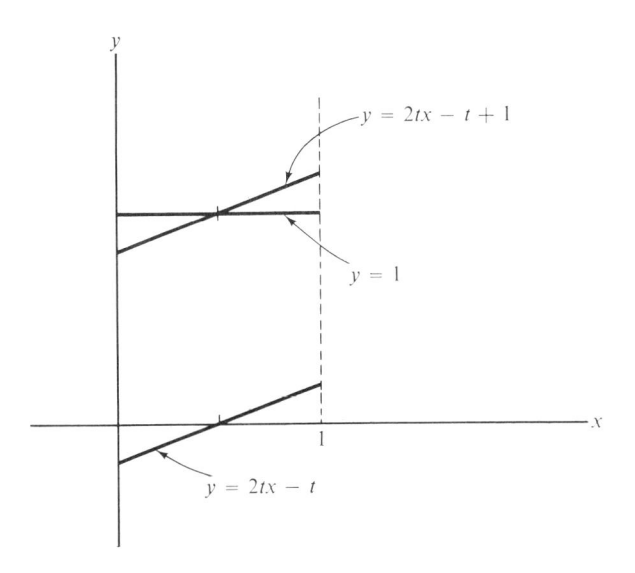

FIGURE 14-3

Already we have studied the composition and inversion of linear transformations and we have identified these operations with the multiplication and inversion of matrices. Similarly, the new operations we are about to define correspond to matrix operations.

14-3 Definition

Let S and T be linear transformations of V into W, and let a be any scalar. The transformations $S + T$ and aT, each of V into W, are defined by

$$(S + T)(\alpha) = S(\alpha) + T(\alpha) \quad \text{for all } \alpha \varepsilon V$$
$$(aT)(\alpha) = aT(\alpha) \quad \text{for all } \alpha \varepsilon V$$

Example 5. Let S and T be mappings of R^2 into R^3, with $S(x, y) = (x, y, x + y)$ and $T(x, y) = (x, x + y, x + y + z)$. Then

Lessons in Linear Algebra

$$(S + T)(x, y) = (2x, x + 2y, 2x + 2y + z)$$

and

$$(aT)(x, y) = (ax, ax + ay, ax + ay + az)$$

Consider the vector $\alpha = (1, 1)$. S sends α into $(1, 1, 2)$, T sends α into $(1, 2, 3)$, and $S + T$ sends α into $(2, 3, 5)$. Also, $3T$ sends α into $(3, 6, 9)$.

Actually we did not use the operations in V in defining $S + T$ and aT. The new operations could be defined for any mapping of an arbitrary set into W. However, we will restrict our attention to the case where V and W are both linear spaces over the same field.

In Example 5, $S + T$ and aT both turned out to be linear, a general result.

We will use $L(V, W)$ to denote the set of all linear transformations of V into W.

14-4 Theorem

$L(V, W)$ is itself a linear space, under the operations of addition and scalar multiplication defined in Definition 14-3.

Proof: First we need to show $L(V, W)$ is closed. Suppose S, T are linear. Then $S + T$ is linear because for all $\alpha_1, \alpha_2 \in V$ and scalars c_1, c_2,

$$(S + T)(c_1\alpha_1 + c_2\alpha_2) = S(c_1\alpha_1 + c_2\alpha_2) + T(c_1\alpha_1 + c_2\alpha_2)$$
(Definition 14-3)
$$= c_1S(\alpha_1) + c_2S(\alpha_2) + c_1T(\alpha_1) + c_2T(\alpha_2)$$
[linearity]
$$= c_1(S + T)(\alpha_1) + c_2(S + T)(\alpha_2).$$
(Definition 14-3)

Similarly aT is linear.

It is easy to verify that $L(V, W)$ satisfies all of the other axioms for a linear space. For example, $L(V, W)$ is associative because from Theorem 14-3 and the associative law in W,

$$[(S + T) + U](\alpha) = (S + T)(\alpha) + U(\alpha)$$
$$= S(\alpha) + T(\alpha) + U(\alpha)$$
$$= [S + (T + U)](\alpha)$$

and, hence, $(S + T) + U = S + (T + U)$.

The zero transformation is the one which maps each element of V into 0, and the negative of T is defined by $(-T)(\alpha) = -T(\alpha)$.

It is left for the reader to check that the other axioms for a linear space are satisfied. ∎

The Linear Problem — More Operations on Linear Transformations

In case V and W are finite-dimensional, then $L(V, W)$ is isomorphic to a linear space of matrices. In preparation for the proof of this, we note that the set of all m by n matrices form a vector space, because the set is closed under addition, scalar multiplication, etc. In fact an m by n matrix is an mn-tuple.

14-5 Theorem. Let $\dim V = n$, $\dim W = m$. Then $L(V, W)$ is isomorphic to $\mathfrak{M}$, the set of all m by n matrices.

Proof: We represent each transformation in $L(V, W)$ relative to the bases $\{\alpha_1, \ldots, \alpha_n\}$ and $\{\beta_1, \ldots, \beta_m\}$. The matrix of $T + S$ is $[t_{ij}] + [s_{ij}]$, the matrix of T added to the matrix of S, for

$$(T + S)(\alpha_j) = T(\alpha_j) + S(\alpha_j) \quad \text{[linearity]}$$

$$= \sum_{i=1}^{m} t_{ij}\beta_i + \sum_{i=1}^{m} s_{ij}\beta_i$$

[definition of matrix of a transformation]

$$= \sum_{i=1}^{m} (t_{ij} + s_{ij})\beta_i$$

Similarly

$$(aT)(\alpha_j) = aT(\alpha_j)$$
$$= a \sum t_{ij}\beta_i$$
$$= \sum at_{ij}\beta_i$$

and the matrix of aT is a times the matrix of T.

Let $\phi : L(V, W) \to \mathfrak{M}$ by $\phi(T) = [t_{ij}]$, where $[t_{ij}]$ is the matrix of T. Then ϕ is 1-1, onto and $\phi(aT_1 + bT_2) = a\phi(T_1) + b\phi(T_2)$. Hence, $L(V, W)$ is isomorphic to $\mathfrak{M}$. ∎

Now we are in a position to use the two algebras (that of $L(V, W)$ and that of $\mathfrak{M}$) interchangeably.

Example 6. Let S and T be mappings of R^2 into R^2, with $S(x, y) = (x + 2y, 3x - y)$, and $T(x, y) = (x - y, y)$. Then $ST(x, y) = (x + y, 3x - 4y)$, $TS(x, y) = (-2x + 3y, 3x - y)$, and $(ST + TS)(x, y) = (-x + 4y, 6x - 5y)$.

Lessons in Linear Algebra

We may use matrix algebra to cover the same ground. Relative to the standard basis the matrices representing S and T are, respectively.

$$A = \begin{bmatrix} 1 & 2 \\ 3 & -1 \end{bmatrix} \quad \text{and} \quad B = \begin{bmatrix} 1 & -1 \\ 0 & 1 \end{bmatrix}$$

Then the other transformations are represented by

$$AB = \begin{bmatrix} 1 & 1 \\ 3 & -4 \end{bmatrix}, \quad BA = \begin{bmatrix} -2 & 3 \\ 3 & -1 \end{bmatrix}, \quad AB + BA = \begin{bmatrix} -1 & 4 \\ 6 & -5 \end{bmatrix}$$

Also, let us evaluate $T^2 - 2T + I$, where I represents the identity mapping.

$$\begin{aligned}(T^2 - 2T + I)(x, y) &= T(T(x, y)) - 2T(x, y) + (x, y) \\ &= T(x - y, y) - 2(x - y, y) + (x, y) \\ &= (x - 2y, y) - (2x - 2y, 2y) + (x, y) \\ &= (0, 0)\end{aligned}$$

Therefore, $T^2 - 2T + I = 0$, the zero mapping, a fact that can be checked by observing that the matrix B of T satisfies $B^2 - 2B + I = 0$.

Example 7. Let S and T be linear transformations satisfying $ST = TS$. We will prove that $(ST)^n = S^n T^n$ for each positive integer n. The statement holds for $n = 1$. Also, $(ST)^2 = STST = SSTT = S^2 T^2$. We use induction and suppose k is such that $(ST)^k = S^k T^k$. Then $(ST)^{k+1} = S(TS)^k T = SS^k T^k T = S^{k+1} T^{k+1}$.

The proof above remains unchanged if S, T are considered to be n by n matrices satisfying $ST = TS$. Indeed any fact about matrix algebra corresponds to a fact about the algebra of linear transformations, because of Theorem 14-5. You are asked to use this idea to prove the following.

14-6 Theorem The solution space of a system of homogeneous linear equations in n unknowns has dimension $n - r$, where r is the rank of the matrix of the system.

Summary of Topics in Lesson 14

The linear problem. General results, including facts about the homogeneous and nonhomogeneous problems.

The Linear Problem — More Operations on Linear Transformations

Illustrative examples involving systems of linear equations, differential equations, and function spaces, with each example interpreted as a linear problem.

The system $L(V, W)$ of linear transformations of V into W. The sum $S + T$ of two linear transformations, and the scalar multiple aT of a linear transformation.

The system $L(V, W)$ is a linear space.

If V and W are finite-dimensional, then $L(V, W)$ is isomorphic to an algebra of m by n matrices.

Exercises

1. Solve the system
$$2x + y - z + w = 5$$
$$x - y + z - w = 1$$
and display the general solution as the sum of a particular solution and a solution of the corresponding homogeneous problem.

2. Of use to engineers and physicists is the differential equation $y'' + a^2 y = f(x)$. The corresponding homogeneous problem $y'' + a^2 y = 0$ has the general solution $y = c_1 \cos ax + c_2 \sin ax$.
   (a) Solve $y'' + 9y = e^x$. Hint: Find a particular solution of the form Ae^x.
   (b) Solve $y'' + 9y = x^2 + 1$ by finding a solution of the form $Ax^2 + Bx + C$.
   (c) Given that α_1 is a solution of $T(\alpha) = \beta_1$, and α_2 is a solution of $T(\alpha) = \beta_2$, what can you say about $\alpha_1 + \alpha_2$?
   (d) Solve $y'' + 9y = e^x + x^2 + 1$.

3. Let V be the function space spanned by $\{e^x, e^{-x}\}$, and let $T : V \to R$ by $T(f(x)) = \int_0^1 f(x)\, dx$. Find the functions of V for which
   (a) $T(f(x)) = 0$.
   (b) $T(f(x)) = e - 1$.

4. Prove part (e) of Theorem 14-2.

5. (a) Given that $(1, 2, 3)$ and $(3, 1, 1)$ are solutions of a system of homogeneous linear equations, find a solution of the form $(-2, y, z)$.
   (b) Given that $(1, 2, 3)$ and $(3, 1, 1)$ are solutions of a nonhomogeneous system, find a solution of the form $(-2, y, z)$.

6. Let $S, T: R^2 \to R^2$ with $S(x, y) = (2x, x + y)$, $T(x, y) = (x - 2y, x)$. Find rules for S^2, T^2, ST, TS, $(ST)^2$, S^2T^2, $ST - TS$, S^{-1}, T^{-1}, $(ST)^{-1}$, $T^{-1}S^{-1}$, and $S^{-1}T^{-1}$. (You may use matrices if you like.)

7. Let S, T map $V \to W$ and let $R: W \to X$. Show that $R(S + T) = RS + RT$. Also, state and prove a similar right distributive law.

8. If S, T are invertible elements of $L(V, V)$, explain why
    (a) $(ST)^{-1} = T^{-1}S^{-1}$.
    (b) $(S^{-1})^n = (S^n)^{-1}$ if n is a positive integer.
    (c) $S^m S^n = S^{m+n}$ for all integers m, n.
    (d) $(S^m)^n = S^{mn}$ for all integers m, n.

9. Let V be the space of twice-differentiable functions; let D, T be mappings of V into V with $D(f(x)) = f'(x)$; and let $T(f(x)) = (x + 1)f(x)$.
    (a) Find $DT - TD$, show $DT^2 + T^2D = 2T$, and develop an expression for $DT^n - T^nD$, where $n > 0$. Prove that your conjecture is correct.
    (b) If A, B are square matrices it is impossible that $AB - BA = I$, because the trace of $AB - BA$ is zero. How can this be resolved with part (a)?

10. What is the dimension of the linear space consisting of all m by n matrices? Find a basis.

11. Let V be the space of 2 by 2 matrices and let
$$E_1 = \begin{bmatrix} 1 & 0 \\ 0 & 0 \end{bmatrix}, \quad E_2 = \begin{bmatrix} 0 & 1 \\ 0 & 0 \end{bmatrix}, \quad E_3 = \begin{bmatrix} 0 & 0 \\ 1 & 0 \end{bmatrix}, \quad E_4 = \begin{bmatrix} 0 & 0 \\ 0 & 1 \end{bmatrix}$$
Show that the matrix A commutes with each matrix of V if and only if A commutes with each E_i.

12. Describe the matrices which commute with every n by n matrix.

13. Let A be an n by n matrix. Show that there is a polynomial p such that $p(A)$ is the zero matrix. To illustrate what this means, if $p(x) = x^2 - 4x + 3$ and
$$A = \begin{bmatrix} 1 & 0 \\ 0 & 3 \end{bmatrix}$$
then $p(A) = A^2 - 4A + 3I = 0$. Hint: Consider the set A^i, where $i = 0, 1, \ldots, n^2$, as a set of vectors in the space of n by n matrices.

14. Let $T: P_1 \to P_1$ by $T(p(x)) = 2p'(x) + p(x)$. (P_1 is the space of polynomials of degree less than or equal to 1.) Find the rule for $2T - T^2$.

The Linear Problem — More Operations on Linear Transformations

15. Let
$$A = \begin{bmatrix} 1 & 1 \\ 0 & 2 \end{bmatrix}$$
and compute A^2, A^3, A^4. Find a fourth degree polynomial $p(x)$ such that $p(A) = 0$. Factor $p(x)$ and determine if $p(x)$ has a factor $q(x)$ for which $q(A) = 0$.

16. Prove Theorem 14-6.

Remarks on Exercises

1. $(2, 1, 0, 0) + a(0, 1, 1, 0) + b(0, -1, 0, 1)$.

2. (a) $y = \dfrac{1}{10} e^x + c_1 \cos 3x + c_2 \sin 3x$, ($c_1$, c_2 arbitrary).

   (b) $y = \dfrac{1}{9} x^2 + \dfrac{7}{81} + c_1 \cos 3x + c_2 \sin 3x$, ($c_1$, c_2 arbitrary),

   (c) $\alpha_1 + \alpha_2$ is a solution of $T(\alpha) = \beta_1 + \beta_2$.

   (d) $\dfrac{1}{10} e^x + \dfrac{1}{9} x^2 + \dfrac{7}{81} + c_1 \cos 3x + c_2 \sin 3x$, ($c_1$, c_2 arbitrary), obtained by adding (a) and (b).

3. (a) $a(e^x - e^{1-x})$, a arbitrary.
   (b) $a(e^x - e^{1-x}) + e^x$.

4. $T(\sum c_i \alpha_i) = \beta \Leftrightarrow \sum c_i T(\alpha_i) = \beta$ (linearity)
   $\Leftrightarrow \sum c_i \beta = \beta$ (each $T(\alpha_i) = \beta$ by hypothesis)
   $\Leftrightarrow \sum c_i = 1$

5. (a) $(1, 2, 3) - (3, 1, 1) = (-2, 1, 2)$, a solution. The answer is not unique.

   (b) $\dfrac{5}{2}(1, 2, 3) - \dfrac{3}{2}(3, 1, 1) = \left(-2, \dfrac{7}{2}, 6\right)$. The answer is unique.

6.

Transformation	Image of (x, y)	Matrix relative to the standard basis
S	$(2x,\ x + y)$	$\begin{bmatrix} 2 & 0 \\ 1 & 1 \end{bmatrix}$
T	$(x - 2y,\ x)$	$\begin{bmatrix} 1 & -2 \\ 1 & 0 \end{bmatrix}$
S^2	$(4x,\ 3x + y)$	$\begin{bmatrix} 4 & 0 \\ 3 & 1 \end{bmatrix}$
T^2	$(-x - 2y,\ x - 2y)$	$\begin{bmatrix} -1 & -2 \\ 1 & -2 \end{bmatrix}$

Lessons in Linear Algebra

ST	$(2x - 4y, 2x - 2y)$	$\begin{bmatrix} 2 & -4 \\ 2 & -2 \end{bmatrix}$
TS	$(-2y, 2x)$	$\begin{bmatrix} 0 & -2 \\ 2 & 0 \end{bmatrix}$
$(ST)^2$	$(-4x, -4y)$	$\begin{bmatrix} -4 & 0 \\ 0 & -4 \end{bmatrix}$
S^2T^2	$(-4x - 8y, -2x - 8y)$	$\begin{bmatrix} -4 & -8 \\ -2 & -8 \end{bmatrix}$
$ST - TS$	$(2x - 2y, -2y)$	$\begin{bmatrix} 2 & -2 \\ 0 & -2 \end{bmatrix}$
S^{-1}	$\left(\frac{1}{2}x, -\frac{1}{2}x + y\right)$	$\frac{1}{2}\begin{bmatrix} 1 & 0 \\ -1 & 2 \end{bmatrix}$
T^{-1}	$\left(y, -\frac{1}{2}x + \frac{1}{2}y\right)$	$\frac{1}{2}\begin{bmatrix} 0 & 2 \\ -1 & 1 \end{bmatrix}$
$(ST)^{-1}$	$\left(-\frac{1}{2}x + y, -\frac{1}{2}x + \frac{1}{2}y\right)$	$\frac{1}{2}\begin{bmatrix} -1 & 2 \\ -1 & 1 \end{bmatrix}$
$T^{-1}S^{-1}$	$\left(-\frac{1}{2}x + y, -\frac{1}{2}x + \frac{1}{2}y\right)$	$\frac{1}{2}\begin{bmatrix} -1 & 2 \\ -1 & 1 \end{bmatrix}$
$S^{-1}T^{-1}$	$\left(\frac{1}{2}y, -\frac{1}{2}x\right)$	$\frac{1}{2}\begin{bmatrix} 0 & 1 \\ -1 & 0 \end{bmatrix}$

7. For $\alpha \varepsilon V$,

$$\begin{aligned}
[R(S + T)](\alpha) &= R[(S + T)(\alpha)] & \text{(composition)} \\
&= R[S(\alpha) + T(\alpha)] & \text{(definition of } S + T) \\
&= R(S(\alpha)) + R(T(\alpha)) & \text{(linearity of } R) \\
&= (RS)(\alpha) + (RT)(\alpha) & \text{(composition)} \\
&= (RS + RT)(\alpha) & \text{(definition of sum of transformations)}
\end{aligned}$$

Therefore,
$$R(S + T) = RS + RT$$

8. (b) Use induction. For $n = 1$, clearly $(S^{-1})^n = (S^n)^{-1}$. Let k be such that $(S^{-1})^k = (S^k)^{-1}$. Then

$$(S^{-1})^{k+1} = (S^{-1})^k(S^{-1}) = (S^k)^{-1}(S^{-1}) = (SS^k)^{-1} = (S^{k+1})^{-1}$$

(c) We may use induction if m, n are positive. For another case, let m, n be negative with $m = -p$, $n = -q$. Then $S^m S^n = S^{-p}S^{-q} = (S^{-1})^p(S^{-1})^q = (S^{-1})^{p+q} = S^{-p-q} = S^{m+n}$. This leaves the case where one exponent is positive, the other negative.

(d) For $n > 0$ use induction. For $n = -p$, $S^{mn} = (S^m)^{-p} = [(S^m)^{-1}]^p = (S^{-1})^{mp} = S^{-mp} = S^{mn}$.

The Linear Problem — More Operations on Linear Transformations

9. (a) $DT - TD = I$, the identity mapping, $DT^2 - T^2D = 2T$, $DT^3 - T^3D = 3T^2$, and we conjecture that $DT^n - T^nD = nT^{n-1}$. We use induction to establish the last equation. With the induction hypothesis $DT^k - T^kD = kT^{k-1}$, we have
    $$\begin{aligned} DT^{k+1} - T^{k+1}D &= DT^{k+1} - T^kDT + T^kDT - T^{k+1}D \\ &= (DT^k - T^kD)T + T^k(DT - TD) \\ &= kT^{k-1}T + T^kI \\ &= (k+1)T^k \end{aligned}$$

    (b) The transformations T and D used in part (a) act on an infinite-dimensional vector space. If D, T could be represented by square matrices, then we would have a contradiction.

10. mn. There is a "standard" basis. See Exercise 11.

11. Use the fact that the general matrix $B = \begin{bmatrix} b_1 & b_2 \\ b_3 & b_4 \end{bmatrix}$ can be written $\sum_{i=1}^{4} b_i E_i$. If A commutes with every matrix, obviously A commutes with each E_i. Conversely, suppose $AE_i = E_i A$ for each i. Then $AB = A \sum b_i E_i = \sum b_i AE_i = \sum b_i E_i A = BA$.

12. Let $A = [a_{ij}]_{n,n}$. If A commutes with every $n \times n$ matrix, then A must commute with the matrix E which has only one nonzero element $e_{rs} = 1$. Equating the rs-elements of EA and AE gives
    $$\sum_{k=1}^{n} e_{rk}a_{ks} = \sum_{h=1}^{n} a_{rh}e_{hs} \quad \text{or} \quad a_{ss} = a_{rr}$$
    so the elements along the main diagonal are all equal. Equating the ts-elements of EA and AE ($t \neq r$), gives
    $$\sum_{k=1}^{n} e_{tk}a_{ks} = \sum_{h=1}^{n} a_{th}e_{hs} \quad \text{or} \quad 0 = a_{tr}$$
    Therefore, A must be a multiple of I.

13. The set $\{A^i\}$ consists of $n^2 + 1$ vectors in a space of dimension n^2. Hence, they are linearly dependent and there exists a nontrivial linear relation
    $$c_0 I + c_1 A + c_2 A^2 + \ldots + c_m A^m = 0, \quad m = n^2$$

14. $2T - T^2 = I$, the identity mapping on P_1.

15. One such polynomial is $p(x) = x^4 - 3x^3 + 2x^2$. In fact, let $q(x) = x^2 - 3x + 2$ and we have $q(A) = 0$.

16. Use Theorem 13-3: The rank of T plus the nullity of T equals the dimension of the domain of T.

15

DETERMINANTS

In order to continue efficiently our study of matrices and linear transformations we need to become familiar with determinants. Accordingly, the next two lessons are used to define and illustrate determinants, to list some of the properties of determinants, to practice the evaluation of determinants, and to show how determinants are used in other areas, especially in linear algebra.

The determinant of a square matrix is a scalar defined in a moderately complicated fashion. First we define *inversion*, then *determinant*.

15-1 Definition

Let $p_1, p_2, p_3, \ldots, p_n$ be a permutation or arrangement of distinct positive integers. Each pair of integers p_i, p_j where $i < j$ and $p_i > p_j$ forms an *inversion*.

For example, consider the arrangement 24316. Note that $p_1 = 2$, $p_2 = 4$, $p_3 = 3$, $p_4 = 1$, $p_5 = 6$. Since $1 < 4$ and $p_1 > p_4$, the pair 2, 1 forms an inversion. Since $2 < 3$ and $p_2 > p_3$, the pair 4, 3 forms an inversion. Similarly the pairs (4, 1) and (3, 1) form inversions. The permutation 24316 has exactly four inversions.

15-2 Definition

Let $A = [a_{ij}]$ be a square matrix of order n. The *determinant* of A, denoted det A or $|A|$, is the sum of all terms of the form

$$(-1)^p a_{1i_1} a_{2i_2} \ldots a_{ni_n}$$

where the row indices are in natural order $1, 2, \ldots, n$ and p is number of inversions presented by the permutation $i_1 i_2 \ldots i_n$ of the column indices.

The sign $(-1)^p$ is positive if the column indices present an even number p of inversions. The sign is negative if the number of such inversions is odd.)

Determinants

As an illustration, consider the matrix

$$A = \begin{bmatrix} a_{11} & a_{12} & a_{13} \\ a_{21} & a_{22} & a_{23} \\ a_{31} & a_{32} & a_{33} \end{bmatrix}$$

and its determinant

$$\det A = |A| = \begin{vmatrix} a_{11} & a_{12} & a_{13} \\ a_{21} & a_{22} & a_{23} \\ a_{31} & a_{32} & a_{33} \end{vmatrix}$$

The permutations of the set 1, 2, 3 are six in number: 123, 132, 213, 231, 312, 321, and hence the terms of $|A|$ are $a_{11}\,a_{22}\,a_{33}$, $a_{11}\,a_{23}\,a_{32}$, $a_{12}\,a_{21}\,a_{33}$, $a_{12}\,a_{23}\,a_{31}$, $a_{13}\,a_{21}\,a_{32}$, $a_{13}\,a_{22}\,a_{31}$. In order to determine the signs of these terms, we count inversions. In the permutation 123 there are no inversions since 1 precedes 2, 1 precedes 3, and 2 precedes 3; that is, any pair of integers are in natural order. In the permutation 132, 1 precedes 3, 1 precedes 2, and 3 precedes 2. Therefore, the permutation 132 presents one inversion.

We conclude that the sign of $a_{11}\,a_{22}\,a_{33}$ is positive, while the sign of $a_{11}\,a_{23}\,a_{32}$ is negative. Similarly the signs of the other four terms are established, and

$$|A| = a_{11}\,a_{22}\,a_{33} + a_{12}\,a_{23}\,a_{31} + a_{13}\,a_{21}\,a_{32} - a_{13}\,a_{22}\,a_{31}$$
$$- a_{11}\,a_{23}\,a_{32} - a_{12}\,a_{21}\,a_{33}$$

It is impractical to use directly the definition in evaluating a determinant. The nth order determinant has $n!$ terms. It would be fairly tedious for one to list the terms in a determinant of order greater than four, and also it would be difficult for one to count inversions in order to determine the sign of each term. It is practical to develop rules which facilitate the evaluation of determinants.

There are several simple rules for evaluating the third order determinant. One is as follows: Take the determinant $|A|$ and copy the first two columns after the third column.

$$\begin{vmatrix} a_{11} & a_{12} & a_{13} \\ a_{21} & a_{22} & a_{23} \\ a_{31} & a_{32} & a_{33} \end{vmatrix} \begin{matrix} a_{11} & a_{12} \\ a_{21} & a_{22} \\ a_{31} & a_{32} \end{matrix}$$

The terms given in Definition 15-2 can be identified with the lines drawn in the preceding array. The positive terms correspond to the lines sloping downward to the right.

Lessons in Linear Algebra

Example 1. Evaluate the determinant of the matrix

$$A = \begin{bmatrix} 1 & 2 & -1 \\ 3 & 0 & 2 \\ 1 & 4 & -1 \end{bmatrix}$$

Solution:

$$\begin{vmatrix} 1 & 2 & -1 \\ 3 & 0 & 2 \\ 1 & 4 & -1 \end{vmatrix} \begin{matrix} 1 & 2 \\ 3 & 0 \\ 1 & 4 \end{matrix}$$

$$|A| = 0 + 4 - 12 + 0 - 8 + 6 = -10$$

For the second order determinant,

$$\begin{vmatrix} a & b \\ c & d \end{vmatrix} = ad - bc$$

In order to evaluate with efficiency determinants of higher order, other procedures are necessary. You can see how difficult it would be to list terms and count inversions for a determinant of order 4 or 5, because the determinant of order n has $n!$ terms.

Here are some properties which will be useful in the expansion of determinants. In practice, it is usually much better to use these properties than to apply directly the definition of a determinant.

15-3 Properties of Determinants

P1. $|A^T| = |A|$.

P2. If A contains a row (or column) of zeros, then $|A| = 0$.

P3. Let B be the matrix obtained by interchanging two rows (or columns) of A. Then $|B| = -|A|$.

P4. If A contains two identical rows (or columns), then $|A| = 0$.

P5. Let B be the matrix obtained by multiplying a row (or column) of A by r. Then $|B| = r|A|$.

P6. Let A, B, C be identical except in one row, for instance the kth row, and let the kth row of A be the sum of the kth rows of B and C. Then $|A| = |B| + |C|$. (A similar statement holds if the word *row* is replaced by the word *column*.)

P7. The value of the determinant of a matrix is not changed if to any column (row) is added r times any other column (row).

Proofs: In the proofs the determinant is considered to be

$$\sum (-1)^p a_{1i_1} a_{2i_2} \ldots a_{ni_n}$$

where p is the number of inversions in $i_1 i_2 \ldots i_n$.

P1. The general term $(-1)^p a_{1i_1} a_{2i_2} \ldots a_{ni_n}$ of $|A|$ has a sign determined by the number of inversions in the column indices. Also $\pm a_{1i_1} a_{2i_2} \ldots a_{ni_n}$ is a term of $|A^T|$, but the sign of this term is determined by the number of inversions presented by the first indices of the elements $a_{1i_1}, a_{2i_2}, \ldots$ after they have been rearranged so that the indices $i_1, i_2, \ldots, i_n$ are in natural order. Consider the effect of this rearrangement on the number of inversions. In the term $a_{1i_1} a_{2i_2} \ldots a_{ni_n}$, consider any two elements a_{ki_k} and a_{mi_m}, where $k < m$. If i_k and i_m are in natural order, then the rearrangement of the factors of the term $a_{1i_1} a_{2i_2} \ldots a_{ni_n}$ leaves k, m in natural order. If i_k and i_m present an inversion, then the rearrangement leads to the inversion m, k. Therefore, the terms of $|A|$ are the same as those of $|A^T|$ and, hence, $|A| = |A^T|$.

The property P1 implies that any statement involving the rows of the general nth order determinant is equivalent to a similar statement involving the columns.

P2. The proof is left for the student.

P3. First it is proved that interchanging two adjacent rows of a matrix changes the sign of its determinant. Consider the term $t = (-1)^p a_{1i_1} a_{2i_2} \ldots a_{ni_n}$. If two adjacent rows of the matrix are swapped, then the new determinant has a term equal to $-t$. Therefore, each term of $|B|$ is the negative of a term of $|A|$, and conversely.

Next, suppose the ith and jth rows are interchanged. The student may verify that this can be accomplished by an odd number of interchanges of adjacent rows. Therefore, interchanging any two rows of A changes the sign of the determinant.

P4. Let the ith and jth rows of A be identical, and let B be the matrix obtained by interchanging the ith and jth rows of A. Since the interchanged rows are identical, $|A| = |B|$. By P3, $|A| = -|B|$. Therefore, $|A| = -|A|$ and $|A| = 0$.

P5. The proof is left for the student.

P6. Take the general elements of A, B, C as a_{ij}, b_{ij}, c_{ij}. By hypothesis $a_{kj} = b_{kj} + c_{kj}$, for $j = 1, 2, \ldots, n$, and $a_{ij} = b_{ij} = c_{ij}$ for all $i \neq k$, $j = 1, \ldots, n$. The general terms of $|A|, |B|, |C|$ are respectively $(-1)^p a_{1i_1} \ldots a_{ki_k} \ldots a_{ni_n}$, $(-1)^p a_{1i_1} \ldots b_{ki_k} \ldots a_{ni_n}$, $(-1)^p a_{1i_1} \ldots c_{ki_k} \ldots a_{ni_n}$. The three terms differ only in the kth factors, and the first term is the sum of the other two. Therefore, $|A| = |B| + |C|$.

Lessons in Linear Algebra

P7. The determinant of A is given by
$$|A| = \sum (-1)^p a_{1i_1} \cdots a_{ni_n}$$

Consider the matrix B obtained by adding to the kth row of A r times the mth row of A.

$$|B| = \sum (-1)^p a_{1i_1} \ldots (a_{ki_k} + r a_{mi_k}) \ldots a_{ni_n}$$
$$= \sum (-1)^p a_{1i_1} \ldots a_{ki_k} \ldots a_{ni_n} + \sum (-1)^p a_{1i_1} \ldots r a_{mi_k} \ldots a_{ni_n}$$
(from P6)
$= |A| + r|C|$ (where C has two equal rows)
$= |A| + 0$ (by P4)
$= |A|$. ■

You may find it helpful to write out a general determinant in order to see clearly the preceding proof.

Example 2. Let
$$A = \begin{bmatrix} 1 & 2 & 3 \\ 4 & -5 & 8 \\ 1 & 4 & 7 \end{bmatrix} \text{ and } B = \begin{bmatrix} 3 & 2 & 1 \\ 8 & -5 & 4 \\ 7 & 4 & 1 \end{bmatrix}$$

Then $|A| = -|B|$, because A and B are the same except for a swap of two columns.
Let
$$C = \begin{bmatrix} 1 & 2 & 3 \\ 10 & 20 & 30 \\ 51 & 47 & -11 \end{bmatrix}$$

By P5 and P4, we have
$$\det C = 10 \begin{vmatrix} 1 & 2 & 3 \\ 1 & 2 & 3 \\ 51 & 47 & -11 \end{vmatrix} = 0$$

In fact P5 and P4 together say that if one row (column) of a matrix is a multiple of another, then the determinant is 0.
Property P6 is illustrated by

$$\begin{vmatrix} 1 & 15 & 3 \\ 2 & 1 & 7 \\ 4 & -5 & 0 \end{vmatrix} = \begin{vmatrix} 1 & 14 & 3 \\ 2 & 2 & 7 \\ 4 & 7 & 0 \end{vmatrix} + \begin{vmatrix} 1 & 1 & 3 \\ 2 & -1 & 7 \\ 4 & -12 & 0 \end{vmatrix}$$

Notice that the three arrays are identical except in the second columns, two of which add up to the other.

Finally

$$\begin{vmatrix} 1 & 2 & 3 \\ -4 & 1 & 5 \\ 2 & 1 & 1 \end{vmatrix} = \begin{vmatrix} 1 & 2 & 3 \\ 0 & 9 & 17 \\ 0 & -3 & -5 \end{vmatrix}$$

because of a double application of P7.

The remainder of this lesson is spent in developing a good procedure for evaluating determinants. We use "cofactors" and the properties of determinants already established.

15-4 Definition

Let $A = [a_{ij}]$ be a square matrix of order n, and let M_{ij} be the matrix obtained from A by deleting the ith row and jth column.

The *cofactor* of a_{ij}, denoted A_{ij}, is defined by

$$A_{ij} = (-1)^{i+j} \det M_{ij}$$

We shall see that det A can be written as a linear combination of any given row (or column). With the ith row we have

$$\det A = a_{i1} A_{i1} + a_{i2} A_{i2} + \ldots + a_{in} A_{in}$$

Therefore, the designation *cofactor* is appropriate.

We begin with a preliminary theorem, designed to prove a more general theorem.

15-5 Lemma

Let $A = [a_{ij}]_{n,n}$. Then $a_{ij} A_{ij}$ is equal to the sum of all terms of det A which contain a_{ij} as a factor.

Proof: First we take the case where $i = j = 1$. The expansion of $a_{11} A_{11}$ is the sum of $(n-1)!$ terms of the form $(-1)^p a_{11} a_{2i_2} a_{3i_3} \ldots a_{ni_n}$. Each of these is a distinct term of det A containing a_{11}. Since there are exactly $(n-1)!$ terms in det A containing a_{11}, the proof is complete. (Note that the number of inversions in 1 i_2 i_3 ... i_n is equal to the number of inversions in i_2 i_3 ... i_n.)

For the other case, let either $i \neq 1$ or $j \neq 1$. The expansion of $a_{ij} A_{ij}$ contains $(n-1)!$ terms, each equal in absolute value to one of the $(n-1)!$ terms of det A containing a_{ij}. It remains only to show that the terms also agree in sign.

Lessons in Linear Algebra

Construct the matrix B from A as follows. Interchange the row containing a_{ij} with the adjacent row above until a_{ij} is in the first row. Then swap adjacent columns until a_{ij} appears in the upper left corner. Call this new matrix $B = [b_{ij}]$, and note that $b_{11} = a_{ij}$. We have already shown that $b_{11} B_{11}$ is the sum of all terms of det B containing b_{11} or a_{ij}. The matrix B was obtained from A by means of $i - 1$ row swaps and $j - 1$ column swaps, and therefore, the terms of det B are equal to $(-1)^{i-1+j-1}$ times the corresponding terms of det A. This shows $a_{ij} A_{ij} = (-1)^{i+j} b_{11} B_{11} = $ the sum of the terms of det A which contain a_{ij} as a factor. ∎

The next theorem is useful in the evaluation of determinants.

15-6 Theorem

Let A be an n by n matrix and choose a row i. Then

(a) $a_{i1} A_{i1} + a_{i2} A_{i2} + \ldots + a_{in} A_{in} = \det A$,
(b) $a_{i1} A_{j1} + a_{i2} A_{j2} + \ldots + a_{in} A_{jn} = 0$, if $i \neq j$.

The two statements can be combined into one:

$$\sum_{k=1}^{n} a_{ik} A_{jk} = \delta_{ij} |A|$$

where the "Kronecker"* delta is defined by

$$\delta_{ij} = \begin{cases} 1 \text{ if } i = j \\ 0 \text{ if } i \neq j \end{cases}$$

Also Theorem 15-6 applies to a column i, with $\sum_{k=1}^{n} a_{ki} A_{kj} = \delta_{ij}|A|$.

Proof: (a) We use Lemma (15-5). The product $a_{i1} A_{i1}$ gives the $(n-1)!$ terms of det A containing a_{i1} as a factor, $a_{i2} A_{i2}$ gives the $(n-1)!$ terms containing a_{i2}, etc. Hence, $\sum_{k} a_{ik} A_{ik}$ gives all the $n(n-1)!$ or $n!$ terms of det A.

(b) The proof is left for you. Compare $\sum_{k} a_{ik} A_{jk}$ ($i \neq j$) with the determinant

---

*The brilliant German mathematician Leopold Kronecker (1823–1891) was primarily an algebraist, showing little interest in geometry, even though he attended the University of Berlin while the great geometer Steiner was there. Kronecker's strongly expressed doubts about the work of contemporary analysts such as Weierstrass has had a good effect on mathematics.

Determinants

$$\begin{vmatrix} a_{11} & a_{12} & \cdots & a_{1n} \\ & \vdots & & \\ (\text{row } i) \quad a_{i1} & a_{i2} & \cdots & a_{in} \\ & \vdots & & \\ (\text{row } j) \quad a_{i1} & a_{i2} & \cdots & a_{in} \\ & \vdots & & \\ a_{n1} & a_{n2} & \cdots & a_{nn} \end{vmatrix}$$

which may be expanded using the cofactors of elements in the jth row. ∎

Next we show how to use Theorem 15-6 in evaluating determinants. The plan is to select a row (or column) of A and then use the cofactors of elements to evaluate det A. Before we select the row we may apply properties P1–P7. Sometimes this makes the computation much easier.

Example 3. Evaluate
$$\begin{vmatrix} 1 & 2 & -1 \\ 3 & 0 & 2 \\ 1 & 4 & -1 \end{vmatrix}$$
by using cofactors of elements in the first row.

Solution:
$$\begin{vmatrix} 1 & 2 & -1 \\ 3 & 0 & 2 \\ 1 & 4 & -1 \end{vmatrix} = (1)\begin{vmatrix} 0 & 2 \\ 4 & -1 \end{vmatrix} - 2\begin{vmatrix} 3 & 2 \\ 1 & -1 \end{vmatrix} + (-1)\begin{vmatrix} 3 & 0 \\ 1 & 4 \end{vmatrix}$$
$$= (1)(-8) - 2(-5) + (-1)(12)$$
$$= -10$$

Example 4. Evaluate
$$|A| = \begin{vmatrix} 3 & -1 & 4 & 1 \\ -1 & 5 & 1 & 4 \\ 4 & 0 & 2 & -1 \\ 2 & 3 & -2 & 1 \end{vmatrix}$$

Lessons in Linear Algebra

Solution: We alter the determinant's form without changing its value by using P7. First, we add to the elements of the first column 3 times the elements of the second column. Then we add to the third column 4 times the second column. Finally, we add to the fourth column 1 times the second column.

$$|A| = \begin{vmatrix} 3 & -1 & 4 & 1 \\ -1 & 5 & 1 & 4 \\ 4 & 0 & 2 & -1 \\ 2 & 3 & -2 & 1 \end{vmatrix} = \begin{vmatrix} 0 & -1 & 0 & 0 \\ 14 & 5 & 21 & 9 \\ 4 & 0 & 2 & -1 \\ 11 & 3 & 10 & 4 \end{vmatrix}$$

Now the determinant is expanded by using the first row according to Theorem 15-6.

$$A = 0\begin{vmatrix} 5 & 21 & 9 \\ 0 & 2 & -1 \\ 3 & 10 & 4 \end{vmatrix} - (-1)\begin{vmatrix} 14 & 21 & 9 \\ 4 & 2 & -1 \\ 11 & 10 & 4 \end{vmatrix} + 0$$

$$+ 0 = \begin{vmatrix} 14 & 21 & 9 \\ 4 & 2 & -1 \\ 11 & 10 & 4 \end{vmatrix}$$

Now we reapply P7, Theorem 15-6, and P5.

$$|A| = \begin{vmatrix} 50 & 39 & 0 \\ 4 & 2 & -1 \\ 27 & 18 & 0 \end{vmatrix} = -(-1)\begin{vmatrix} 50 & 39 \\ 27 & 18 \end{vmatrix} = 9\begin{vmatrix} 50 & 39 \\ 3 & 2 \end{vmatrix}$$
$$= 9(100 - 117) = -153$$

Example 5. We use the determinant

$$|A| = \begin{vmatrix} 1 & 3 & 5 \\ -1 & 4 & 3 \\ 2 & 1 & 2 \end{vmatrix} \text{ to}$$

illustrate part (b) of Theorem 15-6. The cofactors of the elements in the first row are used against the elements in the second row. Then

$a_{21} A_{11} + a_{22} A_{12} + a_{23} A_{13}$

$$= (-1)\begin{vmatrix} 4 & 3 \\ 1 & 2 \end{vmatrix} + 4(-1)\begin{vmatrix} -1 & 3 \\ 2 & 2 \end{vmatrix} + 3\begin{vmatrix} -1 & 4 \\ 2 & 1 \end{vmatrix}$$
$$= (-1)(5) - 4(-8) + 3(-9) = 0$$

In the next lesson we will continue the study of determinants.

Determinants

Summary of Topics in Lesson 15

Definition of determinant. Notation and examples.
Simple rules for evaluating second- and third-order determinants.
Elementary properties of determinants, proved and illustrated.
The expansion of a determinant by means of cofactors.

$$\sum_{k=1}^{n} a_{ik} A_{jk} = \delta_{ij} |A|$$

$$\sum_{k=1}^{n} a_{ki} A_{kj} = \delta_{ij} |A|$$

Exercises

1. Evaluate the following determinants:

$$\begin{vmatrix} 3 & 11 \\ \frac{1}{2} & -4 \end{vmatrix}, \quad \begin{vmatrix} 4 & 0 \\ -1 & \sqrt{3} \end{vmatrix}, \quad \begin{vmatrix} 12 & 25 \\ 36 & 75 \end{vmatrix}$$

2. Evaluate:

(a) $\begin{vmatrix} 2 & 3 & -3 \\ 1 & 2 & 4 \\ -5 & 1 & -2 \end{vmatrix}$ (b) $\begin{vmatrix} 5 & 0 & 1 \\ 2 & 1 & 3 \\ 20 & 10 & 30 \end{vmatrix}$

3. Consider the system of equations

$$a_1 x + a_2 y + a_3 z = 0$$
$$b_1 x + b_2 y + b_3 z = 0$$

and the matrix of the system,

$$A = \begin{bmatrix} a_1 & a_2 & a_3 \\ b_1 & b_2 & b_3 \end{bmatrix}$$

Let A_i ($i = 1, 2, 3$) be the matrix obtained by deleting the ith column of A. Prove that $x = |A_1|, y = -|A_2|, z = |A_3|$ is a solution of the system of equations.

4. Use Exercise 3 in order to find a solution x, y, z (not all zero) of the system

$$2x - y + z = 0$$
$$3x + 5y - 11z = 0$$

179

Lessons in Linear Algebra

5. Find a vector in R^3 orthogonal to each of the vectors (1, 2, 1) and (3, −2, 4). Use Exercise 4.

6. Prove properties P2 and P5.

7. Show that $\begin{vmatrix} xa & xb & xc \\ d & e & f \\ g & h & k \end{vmatrix} = x \begin{vmatrix} a & b & c \\ d & e & f \\ g & h & k \end{vmatrix}$

8. Show that $\begin{vmatrix} a & b & c \\ d & e & f \\ g & h & k \end{vmatrix} = \begin{vmatrix} a & d & g \\ b & e & h \\ c & f & k \end{vmatrix}$

9. Show that $\begin{vmatrix} a & b & c \\ d & e & f \\ g & h & k \end{vmatrix} = -\begin{vmatrix} d & e & f \\ a & b & c \\ g & h & k \end{vmatrix}$

10. Evaluate the determinant $\begin{vmatrix} -1 & 2 & 5 \\ 3 & 1 & 1 \\ 6 & 0 & 2 \end{vmatrix}$

    (a) by using cofactors of elements in the second row.
    (b) by using cofactors of elements in the third column.

11. Prove that $\begin{vmatrix} 2 & 5 & 1 & 2 \\ 3 & 1 & 1 & 0 \\ -1 & -2 & 1 & 4 \\ -4 & -4 & -3 & -6 \end{vmatrix} = 0.$

12. Prove that $\begin{vmatrix} a & b & c & d \\ bcd & acd & abd & abc \\ 1 & 1 & 1 & 1 \\ \overline{a} & \overline{b} & \overline{c} & \overline{d} \\ a & -b & c & -d \end{vmatrix} = 0.$

13. Evaluate $\begin{vmatrix} 1 & -2 & 1 & 3 \\ 1 & -4 & 3 & 1 \\ 1 & -2 & 5 & 6 \\ 2 & 0 & 4 & 7 \end{vmatrix}.$

14. Evaluate $\begin{vmatrix} 2 & -1 & 0 & 3 \\ 1 & 4 & -1 & -1 \\ -3 & 5 & 1 & 4 \\ -6 & 2 & 2 & 2 \end{vmatrix}.$

Determinants

15. Prove that $\begin{vmatrix} x & y & y & y \\ y & y & x & y \\ x & y & x & x \\ x & x & x & y \end{vmatrix} = (x-y)^4$.

16. Prove that $\begin{vmatrix} x & y & 1 \\ x_1 & y_1 & 1 \\ x_2 & y_2 & 1 \end{vmatrix} = 0$ is the equation of the straight line through (x_1, y_1) and (x_2, y_2). Then observe that three points (x_1, y_1) (x_2, y_2), and (x_3, y_3) are collinear if and only if $\begin{vmatrix} x_1 & y_1 & 1 \\ x_2 & y_2 & 1 \\ x_3 & y_3 & 1 \end{vmatrix} = 0$.

17. Prove that $\begin{vmatrix} 1 & 0 & 2 & 0 & 3 \\ 1 & 1 & 2 & 0 & 3 \\ 2 & 1 & 5 & -1 & 6 \\ -4 & 1 & -6 & 3 & -13 \\ 2 & 3 & 3 & 2 & 2 \end{vmatrix} = -19$.

18. Let $a_1 x + b_1 y + c_1 = 0$ and $a_2 x + b_2 y + c_2 = 0$ be two lines. Prove that the lines are parallel if and only if the determinant $\begin{vmatrix} a_1 & b_1 \\ a_2 & b_2 \end{vmatrix} = 0$.

19. The determinant $\begin{vmatrix} 1 & 1 & \cdots & 1 \\ x_1 & x_2 & \cdots & x_n \\ x_1^2 & x_2^2 & \cdots & x_n^2 \\ \cdot & \cdot & & \cdot \\ \cdot & \cdot & & \cdot \\ x_1^{n-1} & x_2^{n-1} & \cdots & x_n^{n-1} \end{vmatrix}$ is called *Vandermonde's determinant*. Show that it is equal to $\prod_{\substack{i,j=1 \\ i>j}}^{n} (x_i - x_j)$, the product of all factors of the form $x_i - x_j$, with $i > j$, and i, j belonging to the set $\{1, 2, \ldots, n\}$.

20. Prove that if A is a skew-symmetric matrix of odd order, then $\det A = 0$. (Recall that skew-symmetry means $A^T = -A$.)

21. Let $A = [a_{ij}]$ be an nth order square matrix with elements which are differentiable functions of x. Prove that
$$\frac{d}{dx} |A| = |A_1| + |A_2| + \ldots + |A_n|$$

Lessons in Linear Algebra

where A_i is the matrix A with each element in the ith row replaced by its derivative.

22. Use Exercise 21 to differentiate

$$f(x) = \begin{vmatrix} x^3 & 3x^2 & 6x & 6 \\ \sin x & \cos x & -\sin x & -\cos x \\ \cos x & -\sin x & -\cos x & \sin x \\ e^x & e^x & e^x & e^x \end{vmatrix}$$

Optional Exercise Involving Complex Numbers

23. For the complex square matrix $A = [a_{ij}]$ let $\overline{A} = [\overline{a_{ij}}]$, the matrix obtained by conjugating each element of A. Prove:
    (a) $\det \overline{A} = \overline{\det A}$.
    (b) if A is Hermitian ($\overline{A^T} = A$), then $\det A$ is real.

Remarks on Exercises

1. $-\dfrac{35}{2}, 4\sqrt{3}, 0$.

2. (a) -103; (b) 0.

4. $(6, 25, 13)$.

5. $(10, -1, -8)$.

10. -32.

13. 60.

14. 0.

16. Argue first that $\begin{vmatrix} x & y & 1 \\ x_1 & y_1 & 1 \\ x_2 & y_2 & 1 \end{vmatrix} = 0$ is a linear equation. Then show (x_1, y_1) and (x_2, y_2) both satisfy the equation.

18. The two lines are parallel if and only if $(a_2, b_2) = m(a_1, b_1)$ for some $m \neq 0$. Now show the determinant $\begin{vmatrix} a_1 & b_1 \\ a_2 & b_2 \end{vmatrix} = 0$ if and only if the rows are proportional.

19. Change the determinant into an equivalent one by: adding to the last row $-x_1$ times the next to the last row, adding to the next to the last row $-x_1$ times the preceding row, etc. This gives

$$\prod_{i=2}^{n}(x_i - x_1)\begin{vmatrix} 1 & 1 & \cdots & 1 \\ 0 & 1 & \cdots & 1 \\ 0 & x_2 & \cdots & x_n \\ \cdot & \cdot & & \cdot \\ \cdot & \cdot & & \cdot \\ \cdot & \cdot & & \cdot \\ 0 & x_2^{n-2} & \cdots & x_n^{n-2} \end{vmatrix}$$

Expand by using the cofactors of the elements in the first column, and use mathematical induction. (The expansion is easy when $n = 2$.) Then the determinant is

$$\prod_{i=2}^{n}(x_i - x_1) \prod_{\substack{i,j=2 \\ i>j}}^{n}(x_i - x_j) = \prod_{\substack{i,j=1 \\ i>j}}^{n}(x_i - x_j)$$

20. $|A| = |A^T| = |-A| = (-1)^n|A| = -|A|$. Hence, $|A| = 0$.

21. $\dfrac{d}{dx}|A| = \dfrac{d}{dx}\sum(-1)^p a_{1i_1} a_{2i_2} \ldots a_{ni_n}$

    $= \sum(-1)^p (a_{1i_1} \ldots a_{ni_n})'$

    $= \sum(-1)^p (a'_{1i_1} a_{2i_2} \ldots a_{ni_n} + a_{1i_1} a'_{2i_2} a_{3i_3} \ldots a_{ni_n}$

    $+ \ldots + a_{1i_1} a_{2i_2} \ldots a'_{ni_n})$

    $= |A_1| + |A_2| + \ldots + |A_n|$.

22. Use the columns of the determinant and apply Exercise 21. The answer is $-2x^3 e^x$.

16

APPLICATIONS OF DETERMINANTS — ORTHOGONAL MATRICES AND ISOMETRIES

Determinants are useful in linear algebra as well as in many other parts of mathematics. We begin this lesson by showing that a set of n vectors in n-space is linearly independent if and only if their determinant is nonzero. Along with this we have that the square matrix A is invertible if and only if det $A \neq 0$. Also, we prove that if A and B are n by n matrices then det $(AB) =$ (det A)(det B).

If A is the matrix of a linear transformation T of $R^n \to R^n$, then det A is the well-known *Jacobian* of T, and in two- or three-space the magnitude of det A gives the factors by which certain areas or volumes are magnified. If det $A = \pm 1$, these areas (volumes) are invariant under T, although distances may change. A negative value for det A indicates that T changes the orientation of any basis.

We will go further and require that $AA^T = I$. Then A is called an *orthogonal* matrix, and T preserves all distances. Such transformations called *isometries* are exceedingly useful.

It is possible to check the linear independence of a set of n vectors in n-space by means of determinants.

16-1 Theorem

The rows (columns) of an n by n matrix are linearly independent if and only if det $A \neq 0$.

Proof: Recall from Lesson 11 that the rows of A are linearly independent if and only if the Hermite normal form (reduced echelon form) of A is the identity matrix I. But changing A to its Hermite normal form B is accomplished by elementary row operations which never change the value of the associated determinant from zero to nonzero, or vice versa.

Therefore, if the rows of A are linearly independent, then det $B \neq 0$ and det $A \neq 0$. Conversely, if the rows of A are linearly dependent, then B has a row of zeros and det $A =$ det $B = 0$. ∎

Applications of Determinants — Orthogonal Matrices and Isometries

The square matrix A is invertible if and only if det $A \neq 0$.

16-2
Corollary

The proof is left for the reader.

Example 1. We check the set of vectors $(1, 3, 5)$, $(-1, 2, 4)$, $(1, 8, 14)$ for linear dependence.
Let
$$A = \begin{bmatrix} 1 & 3 & 5 \\ -1 & 2 & 4 \\ 1 & 8 & 14 \end{bmatrix}$$
Then
$$\det A = \begin{vmatrix} 1 & 3 & 5 \\ 0 & 5 & 9 \\ 0 & 5 & 9 \end{vmatrix} = 0$$
and therefore the set of vectors is linearly dependent. Also the matrix A is singular (noninvertible).

We begin the next topic with another example comparing for matrices A and B the determinants of A, B, AB, and BA.

Example 2. Let $A = \begin{bmatrix} 2 & 1 \\ 3 & 4 \end{bmatrix}$, $B = \begin{bmatrix} -1 & 5 \\ 6 & 2 \end{bmatrix}$. Then $AB = \begin{bmatrix} 4 & 12 \\ 21 & 23 \end{bmatrix}$, $BA = \begin{bmatrix} 13 & 19 \\ 18 & 14 \end{bmatrix}$, and $|A| = 5$, $|B| = -32$, $|AB| = |BA| = -160$, the product of $|A|$ and $|B|$. The result is general.

Let A, B be n by n matrices. Then
$$\det(AB) = (\det A)(\det B)$$

16-3
Theorem

Proof: If either A or B is singular, then AB is also singular, because rank $(AB) \leq \min \{\text{rank } A, \text{rank } B\}$ by Theorem 11-13. So the theorem holds because the determinant of a singular matrix is 0. It remains for us to prove the theorem when A and B are both nonsingular, that is, when $|A| \neq 0$ and $|B| \neq 0$.

First we take care of the case when A is an elementary matrix E. If E is obtained from I by a row swap, then

$$|EB| = -|B| \quad \text{(Swapping two rows changes the sign of the determinant.)}$$
$$= |E|\,|B| \quad [|E| = -1]$$

If E is obtained by multiplying a row of I by $k \neq 0$, then $|EB| = k|B| = |E|\,|B|$, and if E is obtained by adding to one row of I a multiple of another row, then $|EB| = |B| = |E|\,|B|$. Hence, the theorem holds when A is elementary.

Finally, let A be any nonsingular matrix, with $A = E_1 \ldots E_k$, the product of elementary matrices. Then

$$|AB| = |E_1 \ldots E_k B|$$
$$= |E_1|\,|E_2 \ldots E_k B| \quad (|E_1 M| = |E_1|\,|M| \text{ just proved.})$$
$$\vdots$$
$$= |E_1|\,|E_2|\ldots|E_k|\,|B| \quad \text{(same reason)}$$
$$\vdots$$
$$= |E_1 \ldots E_k|\,|B| \quad \text{(same reason)}$$
$$= |A|\,|B|$$

An obvious consequence is that $\det(BA) = (\det A)(\det B)$, because the scalars commute under multiplication. ∎

We introduce next a useful matrix, constructed from the square matrix A by replacing each element a_{ij} by its cofactor A_{ij} and then transposing.

16-4 Definition

Let A be an n by n matrix, and let A_{ij} be the cofactor of a_{ij}. By the *adjoint of A* we mean the transpose of the matrix $[A_{ij}]$. That is,

$$\text{adj } A = \begin{bmatrix} A_{11} & A_{21} & \ldots & A_{n1} \\ A_{12} & A_{22} & \ldots & A_{n2} \\ \vdots & \vdots & & \vdots \\ A_{1n} & A_{2n} & \ldots & A_{nn} \end{bmatrix}$$

We have an immediate result.

Applications of Determinants — Orthogonal Matrices and Isometries

$A(\text{adj } A) = (\det A)I.$

16-5 Theorem

Proof: We use Theorem 15-6, $\sum_{k=1}^{n} a_{ik} A_{jk} = \delta ij |A|$. ∎

Example 3. Let

$$A = \begin{vmatrix} 1 & 4 & 1 \\ 3 & 2 & 4 \\ 5 & 6 & 6 \end{vmatrix}$$

Then $|A| = 4$, and

$$A(\text{adj } A) = \begin{bmatrix} 1 & 4 & 1 \\ 3 & 2 & 4 \\ 5 & 6 & 6 \end{bmatrix} \begin{bmatrix} -12 & -18 & 14 \\ 2 & 1 & -1 \\ 8 & 14 & -10 \end{bmatrix} = \begin{bmatrix} 4 & 0 & 0 \\ 0 & 4 & 0 \\ 0 & 0 & 4 \end{bmatrix} = 4I$$

Since $A(\text{adj } A) = 4I$ it follows that $A^{-1} = (1/4)(\text{adj } A)$, and we have discovered a new method for inverting a matrix. Recall that A is invertible if and only if $\det A \neq 0$ (Corollary 16-2).

If A is invertible, then $A^{-1} = (1/|A|) (\text{adj } A)$.

16-6 Corollary

Proof: Use Theorem 16-5 and multiply by $1/|A|$. ∎

If A represents a linear transformation T of $R^2 \to R^2$ then $|A|$ may be used to measure how T affects areas. (If T were a linear transformation of $R^3 \to R^3$, then $|A|$ would show how T affects volumes.)

Example 4. Let $T: R^2 \to R^2$ with $T(x, y) = (3y, 2x)$. This transformation sends (x, y) into (x', y'), where $x' = 3y$, $y' = 2x$, or equivalently $x = (1/2)y'$, $y = (1/3)x'$.

Consider the unit circle $S = \{(x, y) \mid x^2 + y^2 = 1\}$. The image set is $T(S) = \{(x', y') \mid (1/4)y'^2 + (1/9)x'^2 = 1\}$. We interpret this by saying that the circle $x^2 + y^2 = 1$ goes into the ellipse $(x'^2/9) + (y'^2/4) = 1$.) The area of the circle is π, the area of the ellipse is 6π. (See Figure 16-1.)

The number 6 equals the absolute value of the determinant of the matrix $\begin{bmatrix} 0 & 3 \\ 2 & 0 \end{bmatrix}$, which represents T relative to the standard basis. A

Lessons in Linear Algebra

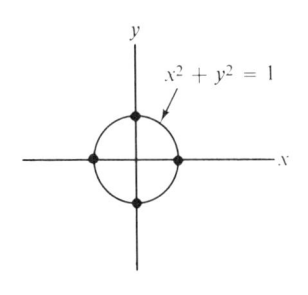

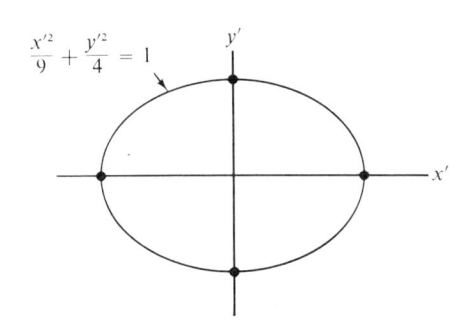

FIGURE 16-1

change in orientation is revealed by the points $(1, 0)$, $(0, 1)$, $(-1, 0)$, $(0, -1)$ and their images $(0, 2)$, $(3, 0)$, $(0, -2)$, $(-3, 0)$.

In order to understand why the change in area is measured by a determinant, let us examine the parallelogram.

16-7 Theorem

Let $\alpha = (a_1, a_2)$ and $\beta = (b_1, b_2)$ be two vectors in the plane. The area of the parallelogram formed by α and β is equal to

(1) the square root of $\begin{vmatrix} \alpha \cdot \alpha & \alpha \cdot \beta \\ \beta \cdot \alpha & \beta \cdot \beta \end{vmatrix}$.

(2) the absolute value of $\begin{vmatrix} a_1 & a_2 \\ b_1 & b_2 \end{vmatrix}$.

Proof: See Figure 16-2. The altitude of the parallelogram is $\beta - [(\beta \cdot \alpha)/(\alpha \cdot \alpha)]\alpha$. So

$$(\text{area})^2 = \|\alpha\|^2 \left\| \beta - \frac{\beta \cdot \alpha}{\alpha \cdot \alpha} \alpha \right\|^2$$

$$= (\alpha \cdot \alpha)\left[\left(\beta - \frac{\beta \cdot \alpha}{\alpha \cdot \alpha} \alpha \right) \cdot \left(\beta - \frac{\beta \cdot \alpha}{\alpha \cdot \alpha} \alpha \right) \right]$$

$$= (\alpha \cdot \alpha)\left[\beta \cdot \beta - 2\frac{\beta \cdot \alpha}{\alpha \cdot \alpha}(\alpha \cdot \beta) + \frac{(\beta \cdot \alpha)^2}{(\alpha \cdot \alpha)^2}(\alpha \cdot \alpha) \right]$$

$$= (\alpha \cdot \alpha)(\beta \cdot \beta) - (\alpha \cdot \beta)^2$$

$$= \begin{vmatrix} \alpha \cdot \alpha & \alpha \cdot \beta \\ \beta \cdot \alpha & \beta \cdot \beta \end{vmatrix}$$

and we have (1). To get (2), we continue and write

Applications of Determinants — Orthogonal Matrices and Isometries

$$(\text{area})^2 = \begin{vmatrix} a_1^2 + a_2^2 & a_1b_1 + a_2b_2 \\ a_1b_1 + a_2b_2 & b_1^2 + b_2^2 \end{vmatrix}$$

$$= \begin{vmatrix} a_1 & a_2 \\ b_1 & b_2 \end{vmatrix} \begin{vmatrix} a_1 & b_1 \\ a_2 & b_2 \end{vmatrix} \qquad \text{[Theorem 16-3]}$$

$$= \begin{vmatrix} a_1 & a_2 \\ b_1 & b_2 \end{vmatrix}^2 \qquad \blacksquare$$

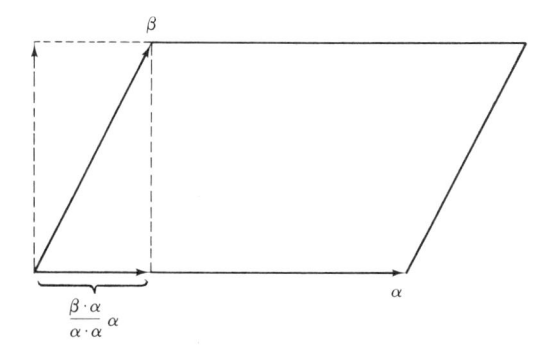

FIGURE 16-2

Let $T: R^2 \to R^2$ be a linear transformation with matrix A relative to the standard basis. Then T multiplies the areas of parallelograms by $|\det A|$.

16-8 Theorem

Proof: The parallelogram determined by $X = (x_1, x_2)$ and $Y = (y_1, y_2)$ goes into the parallelogram determined by $XA^T = (a_{11}x_1 + a_{12}x_2, a_{21}x_1 + a_{22}x_2)$ and $YA^T = (a_{11}y_1 + a_{12}y_2, a_{21}y_1 + a_{22}y_2)$.

Comparing determinants gives

$$\begin{vmatrix} a_{11}x_1 + a_{12}x_2 & a_{21}x_1 + a_{22}x_2 \\ a_{11}y_1 + a_{12}y_2 & a_{21}y_1 + a_{22}y_2 \end{vmatrix} = \begin{vmatrix} x_1 & x_2 \\ y_1 & y_2 \end{vmatrix} \begin{vmatrix} a_{11} & a_{21} \\ a_{12} & a_{22} \end{vmatrix}$$

which shows that the area of the new parallelogram is equal to $|\det A|$ times the area of the old parallelogram. ∎

The extension of the Theorem 16-8 shows that T multiplies all areas by the same number. To prove this, one would make use of the fact that the more general areas are computed as limits of sums of rectangular areas.

We digress to remark that for more general transformations, determinants are useful. For example, let $\phi : (x, y) \to (u, v)$ according to

$u = f(x, y)$, $v = g(x, y)$. Then the *Jacobian** of ϕ is defined as

$$\begin{vmatrix} \frac{\partial f}{\partial x} & \frac{\partial f}{\partial y} \\ \frac{\partial g}{\partial x} & \frac{\partial g}{\partial y} \end{vmatrix}$$

and indicates locally how the transformation magnifies areas.

Whenever the determinant of a linear transformation T of $R^2 \to R^2$ equals ± 1, then areas are invariant under T, but distances may change.

Example 5. Let $T(x, y) = (2x + y, x + y)$. Since $\begin{vmatrix} 2 & 1 \\ 1 & 1 \end{vmatrix} = 1$, T preserves areas, but a look at the elements $\alpha_1 = (1, 0)$ and $\alpha_2 = (0, 1)$ and their images $T(\alpha_1) = (2, 1)$ and $T(\alpha_2) = (1, 1)$ shows that distances are changed, because the distance between α_1 and α_2 is $\sqrt{2}$ while the distance between $T(\alpha_1)$ and $T(\alpha_2)$ is 1.

A transformation which preserves distances is called an *isometry*. We will examine such transformations in R^n and show how they can be represented by very special matrices called *orthogonal* matrices.

16-9 Theorem

Let T be a linear transformation of a Euclidean space V into itself. Then T is an isometry (preserves distances) if and only if T preserves inner products.

Proof: The proof is easy one way. If T preserves inner products, then T preserves norms, for $\|(T\alpha)\| = \sqrt{(T(\alpha), T(\alpha))} = \sqrt{(\alpha, \alpha)} = \|\alpha\|$. The distance from $T(\alpha)$ to $T(\beta)$ is

$$\|T(\alpha) - T(\beta)\| = \|T(\alpha - \beta)\| \qquad \text{(definition of distance, linearity.)}$$
$$= \|\alpha - \beta\| \qquad \text{(T preserves norms.)}$$
$$= \text{distance from } \alpha \text{ to } \beta$$

Conversely, suppose T preserves distances. Then T preserves norms. We seek to conclude that T preserves inner products. To this end we use an identity which gives the inner product in terms of norms.

$$(\alpha, \beta) = \frac{1}{2}[\|\alpha + \beta\|^2 - \|\alpha\|^2 - \|\beta\|^2]$$

---

*Carl Gustav Jacobi (1804–1851), a German mathematician, is remembered especially for his work in curves and surfaces. He is also noted for his contributions in elliptic functions.

Applications of Determinants — Orthogonal Matrices and Isometries

(Check it.)
Using the same identity for the pair $T(\alpha)$, $T(\beta)$, we have

$$(T(\alpha), T(\beta)) = \frac{1}{2}[\|T(\alpha) + T(\beta)\|^2 - \|T(\alpha)\|^2 - \|T(\beta)\|^2]$$

$$= \frac{1}{2}[\|T(\alpha + \beta)\|^2 - \|T(\alpha)\|^2 - \|T(\beta)\|^2]$$

$$= \frac{1}{2}[\|\alpha + \beta\|^2 - \|\alpha\|^2 - \|\beta\|^2] \quad \text{(Norms are preserved.)}$$

$$= (\alpha, \beta) \quad \blacksquare$$

16-10 Definition The real square matrix P is called *orthogonal* if $PP^T = I$, or equivalently if $P^T = P^{-1}$.

For example,

$$\frac{1}{5}\begin{bmatrix} 3 & 4 \\ -4 & 3 \end{bmatrix}$$

is orthogonal. Another example is

$$\frac{1}{3}\begin{bmatrix} 2 & 2 & 1 \\ 1 & -2 & 2 \\ 2 & -1 & -2 \end{bmatrix}$$

The orthogonal matrix has so many nice properties that it is easy to recognize and to use.

16-11 Theorem Let P, Q be n by n matrices. Then

(1) if P and Q are orthogonal, then so are PQ, P^T, and P^{-1}.
(2) the matrix P is orthogonal if and only if the rows (columns) of P are an orthonormal set.
(3) if P is orthogonal, then $\det P = \pm 1$. (If $\det P = 1$, P is called *proper* orthogonal. If $\det P = -1$, P is called *improper* orthogonal.)

Proof: (1) Let P, Q be orthogonal, that is, let $PP^T = QQ^T = I$. Then $(PQ)(PQ)^T = PQQ^TP^T = PIP^T = PP^T = I$, and PQ is orthogonal. The proof that P^T and P^{-1} are also orthogonal is left for you.
(2) Let $P = [p_{ij}]_{n,n}$. Then $PP^T = I$ if and only if the ij-element of PP^T, namely $\sum_k p_{ik} p_{jk} = \delta_{ij}$ (the Kronecker delta), or if and only if the rows of P form an orthonormal set.

Applying the same thinking to the rows of the orthogonal matrix P^T shows that P is orthogonal if and only if its columns form an orthonormal set.

(3) If P is orthogonal, then $PP^T = I$ and $1 = |PP^T| = |P| |P^T| = |P|^2$. ∎

In order to recognize an orthogonal matrix P simply check *one* of the following:

(a) $PP^T = I$.
(b) $P^T P = I$.
(c) $P^T = P^{-1}$.
(d) The rows of P are an orthonormal set.
(e) The columns of P are an orthonormal set.

The converse of part (3) does not hold. It is possible to have $|P| = 1$ for a matrix that is not orthogonal.

We conclude by showing how orthogonal matrices are related to isometries.

16-12 Theorem

Let T be a linear transformation of an n-dimensional Euclidean space V into itself, and let P be the matrix of T relative to an orthonormal basis. Then T is an isometry if and only if P is orthogonal.

Proof: We will show that T preserves inner products if and only if P is orthogonal, and then appeal to Theorem 16-9.

Let us represent typical vectors α, β relative to the given orthonormal basis as $X = (x_1, \ldots, x_n)$, $Y = (y_1, \ldots, y_n)$, respectively. The inner product (α, β) is the same as the dot product XY^T, by Exercise 13, Lesson 7, and the images $T(\alpha)$, $T(\beta)$ are represented by the n-tuples (row vectors) XP^T and YP^T (Theorem 13-4). It follows that

$$(T(\alpha), T(\beta)) = (XP^T)(YP^T)^T = XP^TPY^T$$

and therefore the general inner product is preserved by T if and only if

$$XP^TPY^T = XY^T \quad \text{for all } X, Y$$

or if and only if $P^TP = I$. (See Exercise 10b, Lesson 10.) But this is precisely the condition that P be orthogonal. ∎

Example 6. Let $T: R^2 \to R^2$ with $T(x, y) = (1/5)(4x + 3y, 3x - 4y)$, and the usual dot product. We select the standard basis (actually any

Applications of Determinants — Orthogonal Matrices and Isometries

orthonormal basis would serve) to illustrate the relation between T and its matrix P.

Clearly,

$$P = \frac{1}{5}\begin{bmatrix} 4 & 3 \\ 3 & -4 \end{bmatrix}$$

an orthogonal matrix. Let $\alpha = (5, 3)$, $\beta = (4, 2)$. Then $T(\alpha) = (1/5)(29, 3)$, $T(\beta) = (1/5)(22, 4)$, and $\alpha \cdot \beta = T(\alpha) \cdot T(\beta) = 26$. Also, $\|\alpha\| = \|T(\alpha)\| = \sqrt{34}$.

Example 7. Let V be the subspace of $C(0, 2\pi)$ spanned by the orthonormal basis $\{(1/\sqrt{\pi})\sin x, (1/\sqrt{\pi})\cos x\}$, and let $T = D$, the differentiation operator. Then the matrix of T is

$$\begin{bmatrix} 0 & -1 \\ 1 & 0 \end{bmatrix}$$

an orthogonal matrix. (We have used the inner product $(f, g) = \int_0^{2\pi} f(x)g(x)\,dx$.)

Optional Section on Complex Spaces

With a few modifications, all of the results of the lesson carry over into complex spaces. Theorems 16-1, 16-3, and 16-5 and Corollaries 16-2 and 16-6 are already proved for complex numbers. Theorem 16-9 holds for a complex inner product space V, but the proof requires a change, because the identity $(\alpha, \beta) = (1/2)[\|\alpha + \beta\|^2 - \|\alpha\|^2 - \|\beta\|^2]$ does *not* hold. However, in a complex space the inner product can be expressed in terms of the norm by the identity

$$(\alpha, \beta) = \frac{1}{4}[\|\alpha + \beta\|^2 - \|\alpha - \beta\|^2 + i\|\alpha + i\beta\|^2 - i\|\alpha - i\beta\|^2],$$

and the proof of Theorem 16-9 for a complex inner product space follows.

16-13 Definition

The complex square matrix P is called *unitary* if $PP^* = I$, or equivalently if $P^* = P^{-1}$. (Recall that $P^* = \overline{P^T}$.)

For example,

$$\frac{1}{5}\begin{bmatrix} 3 & 4i \\ 4 & -3i \end{bmatrix}$$

is unitary. It also follows that a real matrix A is unitary if and only if it is orthogonal.

193

Lessons in Linear Algebra

16-14 Theorem

Let P, Q be n by n matrices. Then

(1) if P and Q are unitary, then so are PQ, P^*, P^{-1}, P^T, and $\overline{P}$.
(2) the matrix P is unitary if and only if the rows (columns) of P are an orthonormal set (under the usual inner product $X \cdot Y = XY^*$).
(3) if P is unitary, then $|\det P| = 1$.

The proofs parallel those of Theorem 16-11 and are left for you. Use the properties $(PQ)^* = Q^*P^*$, $(P^T)^{-1} = (P^{-1})^T$, $(\overline{P})^{-1} = \overline{(P^{-1})}$, etc. It should be noted that in (3), $|\det P|$ denotes the modulus of a complex number.

16-15 Theorem

Let T be a linear transformation of an n-dimensional complex space V into itself, and let P be the matrix of T relative to an orthonormal basis. Then T is an isometry if and only if P is unitary.

Proof: We parallel the proof of Theorem 16-12, and show that T preserves inner products if and only if P is unitary. This with Theorem 16-9 would be enough.

Representing typical vectors α, β relative to the given orthonormal basis shows that the inner product $(\alpha, \beta) = XY^*$, where $X = (x_1, \ldots, x_n)$, $Y = (y_1, \ldots, y_n)$. Then

$$(T(\alpha), T(\beta)) = (XP^T)(YP^T)^* = XP^T(P^T)^*Y^*$$

and the general inner product is preserved if and only if

$$X(P^T)(P^T)^*Y^* = XY^* \quad \text{for all } X, Y$$

or if and only if $(P^T)(P^T)^* = I$. But this is equivalent to the statement that P is unitary. ∎

Summary of Topics in Lesson 16

The rows (columns) of a square matrix A are linearly independent if and only if $\det A \neq 0$. This is also the condition that A be invertible.
$\det (AB) = (\det A)(\det B)$.
The adjoint of A. $A(\operatorname{adj} A) = (\det A) I$, which may be used to invert A.
Interpretation of determinants in two-space. The area of a parallelogram as a determinant.
Orthogonal matrices and their properties.

Applications of Determinants — Orthogonal Martices and Isometries

Isometries and orthogonal matrices. A linear transformation T preserves distances if and only if T preserves inner products, or if and only if the matrix of T relative to any orthonormal basis is orthogonal.

Optional for Complex Spaces

Unitary matrices and their properties.

Isometries and unitary matrices. A linear transformation T preserves distances if and only if T preserves inner products, or if and only if the matrix of T relative to any orthonormal basis is unitary.

Exercises

1. Use determinants to check each of the following sets for linear dependence:
   (a) $\{(1, 2, -3), (4, -1, 7), (-1, 7, 2)\}$
   (b) $\{(1, 0, 0, 0), (1, 1, 0, 0), (1, 1, 1, 0), (1, 1, 1, 1)\}$
   (c) $\{(1, 3, 1, 2), (7, 12, 2, 13), (-1, 4, 2, 5), (3, 1, -1, 2)\}$
   (d) $\{(1, 2, 4, 8), (1, 3, 9, 27), (1, 4, 16, 64), (1, 5, 25, 125)\}$
   (A Vandermonde determinant. See Lesson 15, Exercise 19.)

2. Find x so that the set of vectors $\{(1, 1, 2, 0), (2, 1, 1, 0), (-1, 3, 2, 4), (1, 1, 1, x)\}$ is linearly dependent.

3. Let $A = \begin{bmatrix} 1 & 3 & 7 \\ -2 & 1 & 3 \\ 1 & -4 & 3 \end{bmatrix}$.

   (a) Find adj A and check that $A(\text{adj } A) = (\det A) I$.
   (b) Find A^{-1}.

4. (a) Let $A = \begin{bmatrix} 0 & 1 & 1 \\ 1 & 0 & 1 \\ 1 & 1 & 0 \end{bmatrix}$ and $B = \begin{bmatrix} 1 & 1 & 1 \\ 1 & 2 & 3 \\ 1 & 1 & 4 \end{bmatrix}$. Verify that $\det (AB) = (\det A)(\det B)$ by computing the several determinants.

   (b) Prove or disprove: If A and B are nth order matrices, then $\det (A + B) = \det A + \det B$.

5. Prove Corollary 16-2: The nth order matrix A is invertible if and only if $\det A \neq 0$.

6. Let A be an n by n matrix. Prove that $|\text{adj } A| = |A|^{n-1}$. (Do not forget the case where $|A| = 0$.)

7. (a) Given that A is a 2 by 2 matrix, find the relationship between adj (adj A) and A.
   (b) For an n by n invertible matrix, find the relationship between adj (adj A) and A.

8. Show that if P is orthogonal, then P^T and P^{-1} are orthogonal.

9. Let P_1, P_2 be proper orthogonal and let Q_1, Q_2 be improper orthogonal. Classify each of the following as proper orthogonal or improper orthogonal: $P_1 P_2$, $Q_1 Q_2$, $P_1 Q_1$, P_1^{-1}, Q_1^{-1}.

10. Prove or disprove: If P and Q are orthogonal n by n matrices, then $P + Q$ is orthogonal.

11. Construct an orthogonal matrix for which the first row is a multiple of (4, 4, 7).

12. Show that the general second-order proper orthogonal matrix can be written $\begin{bmatrix} \cos \theta & \sin \theta \\ -\sin \theta & \cos \theta \end{bmatrix}$. Similarly, find the general second-order improper orthogonal matrix.

13. (a) Give an example of a matrix A such that det $A = 1$ and yet A is not orthogonal.
    (b) If a symmetric matrix is orthogonal, compare A^{-1} with A. Also simplify A^2, A^n, A^{-n}.

14. The linear transformation $T: R^n \to R^n$ is called *orthogonal* if its matrix relative to some orthonormal basis is orthogonal. Show the following:
    (a) If T is orthogonal, then T sends each orthonormal basis into an orthonormal basis.
    (b) If T sends an orthonormal basis into an orthonormal basis, then T is orthogonal.

15. For a real Euclidean space, prove the identity $(\alpha, \beta) = (1/2)[\|\alpha + \beta\|^2 - \|\alpha\|^2 - \|\beta\|^2]$. (This is a review of an earlier problem. See Lesson 6, Exercise 3(d).)

16. In some books, the rank of the matrix A is defined as the order of the largest nonsingular submatrix of A. Compare this with our fact that the rank of A equals the dimension of the row space (or column space) of A. Illustrate with
$$\begin{bmatrix} 1 & 2 & 3 & 5 \\ 2 & 4 & 6 & 10 \\ 3 & 6 & 9 & 15 \end{bmatrix}$$

Applications of Determinants — Orthogonal Matrices and Isometries

17. Let $S = \{\alpha_1, \ldots, \alpha_r\}$ be a set of vectors of R^n, and let $A = [a_{ij}]$, where $a_{ij} = \alpha_i \cdot \alpha_j$, the dot product of α_i and α_j. Prove that S is linearly dependent if and only if $|A| = 0$. (Do not assume $r = n$.)

18. (a) The linear transformation $T(x, y) = (ax, by)$ sends the unit circle $x^2 + y^2 = 1$ into the ellipse $(x^2/a^2) + (y^2/b^2) = 1$. Find the Jacobian of the transformation and use it to get the area of the ellipse.
    (b) Assuming that the same procedure works for volumes, find the volume of the ellipsoid $(x^2/a^2) + (y^2/b^2) + (z^2/c^2) = 1$.

19. Let $\sum_{j=1}^{n} a_{ij} x_j = c_i$, $i = 1, 2, \ldots, n$ be a system of n linear equations in n unknowns. "Cramer's rule"* states that if $|A| \neq 0$, then the solution of the system is

$$x_k = \frac{D_k}{D} \quad (k = 1, 2, \ldots, n)$$

where $D = |A|$ and D_k is the determinant of the matrix obtained by replacing the kth column of A with the c_i's.
    (a) Illustrate Cramer's rule for the system

$$x + 5y = 3$$
$$x - y = 2$$

    (b) Prove Cramer's rule in the general case.

Exercises on Complex Spaces

20. Verify the identity $(1/4)[\|\alpha + \beta\|^2 - \|\alpha - \beta\|^2 + i\|\alpha + i\beta\|^2 - i\|\alpha - i\beta\|^2] = (\alpha, \beta)$.

21. Complete the proof of Theorem 16-14.

Remarks on Exercises

1. a, b, d independent; c dependent.

2. $x = \frac{1}{2}$.

---

*Gabriel Cramer (1704–1752), a Swiss mathematician, is remembered for his work in determinants. His rule is widely used.

Lessons in Linear Algebra

3. (a) $\text{adj } A = \begin{bmatrix} 15 & -37 & 2 \\ 9 & -4 & -17 \\ 7 & 7 & 7 \end{bmatrix}$. $A(\text{adj } A) = 91I$.

   (b) $A^{-1} = \dfrac{1}{91}(\text{adj } A)$.

4. (a) $\det AB = 6$, $\det A = 2$, $\det B = 3$.
   (b) Not true in general. For example, if $A = B = I$, the statement fails.

5. A is invertible if and only if the Hermite normal form $B = I$. But $|B| = 0$ if and only if $|A| = 0$.

6. By Theorem 16-5, $A(\text{adj } A) = |A|I$. Taking the determinant of each member gives $|A(\text{adj } A)| = |A|^n$, and hence $|A| \, |\text{adj } A| = |A|^n$. If $|A| \neq 0$, we have the desired result $|\text{adj } A| = |A|^{n-1}$. Suppose $|A| = 0$. Then for all $i, j = 1, \ldots, n$ we have $a_{i1} A_{j1} + \ldots + a_{in} A_{jn} = \delta_{ij}|A| = 0$ and hence the rows of adj A are linearly dependent. Then $|\text{adj } A| = 0$, and the statement $|\text{adj } A| = |A|^{n-1}$ holds because $0 = 0$.

7. (a) $\text{adj }(\text{adj } A) = A$.
   (b) $\text{adj }(\text{adj } A) = |A|^{n-2} A$.

9. $P_1 P_2$, $Q_1 Q_2$, P_1^{-1} are proper orthogonal.
   $P_1 Q_1$, Q_1^{-1} are improper orthogonal.

10. This is not true in general. A counterexample is $P = Q = I$.

11. $\begin{bmatrix} \dfrac{4}{9} & \dfrac{4}{9} & \dfrac{7}{9} \\ \dfrac{1}{\sqrt{2}} & -\dfrac{1}{\sqrt{2}} & 0 \\ \dfrac{7}{\sqrt{162}} & \dfrac{7}{\sqrt{162}} & \dfrac{-8}{\sqrt{162}} \end{bmatrix}$. (Answer not unique.)

12. Let $A = \begin{bmatrix} a & b \\ c & d \end{bmatrix}$ be proper orthogonal. Then $A^{-1} = A^T$ gives $\begin{bmatrix} d & -b \\ -c & a \end{bmatrix} = \begin{bmatrix} a & c \\ b & d \end{bmatrix}$ and we have $a = d$, $b = -c$. Also $a^2 + b^2 = 1$. Let θ be such that $a = \cos \theta$, $b = \sin \theta$. Then $A = \begin{bmatrix} \cos \theta & \sin \theta \\ -\sin \theta & \cos \theta \end{bmatrix}$. The general improper orthogonal matrix of second order is $\begin{bmatrix} \cos \theta & \sin \theta \\ \sin \theta & -\cos \theta \end{bmatrix}$.

Applications of Determinants — Orthogonal Matrices and Isometries

13. (b) $A^{-1} = A$. $A^2 = I$. $A^n = A^{-n} = I$ if n is even, $A^n = A^{-n} = A$ if n is odd.

14. (a) If T is orthogonal then by Theorems 16-9 and 16-12 T preserves distances and inner products. Therefore, T sends each orthonormal basis into an orthonormal basis.

    (b) Suppose T sends the orthonormal basis $\{\alpha_1, \ldots, \alpha_n\}$ into the orthonormal basis $\{T(\alpha_1), \ldots, T(\alpha_n)\}$. We will show that T preserves inner products, and then use Theorems 16-9 and 16-12. The inner product $(\sum a_i\alpha_i, \sum b_i\alpha_i) = \sum a_ib_i$. On the other hand, $(T(\sum a_i\alpha_i), T(\sum b_i\alpha_i)) = (\sum a_iT(\alpha_i), \sum b_iT(\alpha_i)) = \sum a_ib_i$.

17. The determinant of A is zero if and only if the columns of A are linearly dependent, or if and only if there exists a nontrivial set of scalars $c_1, \ldots, c_r$ such that $c_1(\alpha_i \cdot \alpha_1) + c_2(\alpha_i \cdot \alpha_2) + \ldots + c_r(\alpha_i \cdot \alpha_r) = 0$, $i = 1, 2, \ldots, r$, or if and only if there are nontrivial c_i's with

    (a) $\alpha_i \cdot (c_1\alpha_1 + \ldots + c_r\alpha_r) = 0$, $\quad i = 1, 2, \ldots, r$

    If the set $\{\alpha_1, \ldots, \alpha_r\}$ is dependent we can choose a nontrivial linear relation $\sum c_i\alpha_i = 0$ and assure (a). Conversely, if (a) holds, then multiply the first equation [of the r equations (a)] by c_1, the second by c_2, etc., and add to get

    $$\left(\sum c_i\alpha_i\right) \cdot \left(\sum c_i\alpha_i\right) = 0$$

    But this implies $\sum c_i\alpha_i = 0$ and $\{\alpha_1, \ldots, \alpha_r\}$ is linearly dependent.

18. (a) Area of the ellipse = (area of circle)(Jacobian)

    $$= (\pi)\begin{vmatrix} a & 0 \\ 0 & b \end{vmatrix} = \pi ab.$$

    (b) $\frac{4}{3}\pi abc$.

19. (b) Consider

    $$a_{11}x_1 + a_{12}x_2 + \cdots + a_{1n}x_n = c_1$$
    $$\vdots$$
    $$a_{n1}x_1 + a_{n2}x_2 + \cdots + a_{nn}x_n = c_n$$

    Multiply the ith equation by A_{ik} and add. By Theorem 15-6, $x_k(a_{1k}A_{1k} + a_{2k}A_{2k} + \cdots + a_{nk}A_{nk}) = c_1A_{1k} + c_2A_{2k} + \cdots + c_nA_{nk}$, or $x_k|A| = D_k$.

17

EIGENVALUES AND EIGENVECTORS

We continue now with our combined study of linear transformations and matrices. Often the properties of a particular transformation T are revealed clearly by a good choice of bases in representing T with a matrix, and with this in mind, we present the concept of eigenvalue and eigenvector of a transformation $T : V \to V$.

After defining and illustrating eigenvalues and eigenvectors, we give a straightforward method of finding the eigenvalues and eigenvectors of certain linear transformations. Also, we develop useful properties related to eigenvalues and eigenvectors.

17-1 Definition

Let $T : V \to V$ be linear. If α is a nonzero vector for which $T(\alpha) = k\alpha$ (k a scalar), then α is an *eigenvector of T* and k is the corresponding *eigenvalue of T*.

(Eigenvalues are sometimes called *characteristic roots*, and eigenvectors are sometimes called *characteristic vectors*.)

It is important to keep in mind that an eigenvalue-eigenvector pair belong to a linear transformation T of a linear space into itself.

Example 1. Let $T : R^2 \to R^2$ according to $T(x, y) = (2x + 3y, 2x + y)$. (See Figure 17-1.)

Note that $T(3, 2) = (12, 8) = 4(3, 2)$. That is, $T(\alpha) = k\alpha$ with $\alpha = (3, 2)$ and $k = 4$. Therefore, $(3, 2)$ is an eigenvector of T corresponding to the eigenvalue 4.

Similarly, from $T(1, -1) = (-1, 1) = -1(1, -1)$ we see that $(1, -1)$ is an eigenvector of T corresponding to the eigenvalue -1.

Eigenvalues and Eigenvectors

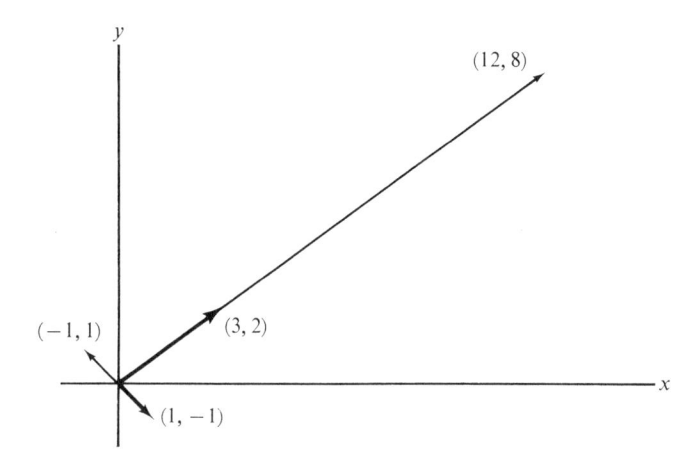

FIGURE 17-1

As we shall see the only eigenvalues of T are 4 and -1. However, there are many eigenvectors of T. For example, $(6, 4)$ is an eigenvector corresponding to the eigenvalue 4.

Example 2. Let D be the differentiation operator, $D(f(x)) = f'(x)$. Then e^{3x} is an eigenvector ("eigenfunction") of D, since $D(e^{3x}) = 3e^{3x}$. The corresponding eigenvalue is 3.

The transformation D^2 has eigenfunction e^{3x} and corresponding eigenvalue 9, since $D^2(e^{3x}) = 9e^{3x}$.

Example 3. Let $T: P_3 \to P_3$ with $T(ax^3 + bx^2 + cx + d) = dx^3 + cx^2 + bx + a$. Then $T(2x^3 + 3x^2 + 3x + 2) = 2x^3 + 3x^2 + 3x + 2$ and $2x^3 + 3x^2 + 3x + 2$ is an eigenvector corresponding to the eigenvalue 1.

Example 4. Let $T: R^2 \to R^2$ with $T(x, y) = (-y, x)$. Since T is a rotation of 90°, T has no eigenvalues or eigenvectors.

On the other hand, for $T: C^2 \to C^2$ by the same rule, we have $T(1, i) = (-i, 1) = -i(1, i)$ and $(1, i)$ is an eigenvector corresponding to the eigenvalue $-i$. So we must keep in mind just what the scalars are.

There is a straightforward method of finding the eigenvalues and eigenvectors of a linear transformation of a finite-dimensional vector space into itself. Relative to a given basis the "eigenvalue" problem $T(\alpha) = k\alpha$ can be expressed in matrix form $AX^T = kX^T$. This is the same as $(A - kI)X^T = 0$ so our problem is to find k and $X \neq 0$ so that X is a solution of the homogeneous system with matrix $A - kI$.

Our procedure is the usual one for homogeneous systems. There is a nontrivial solution X if and only if the matrix $A - kI$ is singular, or if and only if

$$|A - kI| = 0$$

We solve this last equation for k. Any k we find will lead to $X \neq 0$ with $(A - kI)X^T = 0$.

Example 5. We use $T(x, y) = (2x + 3y, 2x + y)$ again. The problem is to find k and $(x, y) \neq (0, 0)$ such that

$$T(x, y) = k(x, y)$$

or

$$(2x + 3y, 2x + y) = k(x, y)$$

or

$$(2 - k)x + 3y = 0$$
$$2x + (1 - k)y = 0$$

There is a nontrivial solution (x, y) if and only if

$$\begin{vmatrix} 2 - k & 3 \\ 2 & 1 - k \end{vmatrix} = 0$$

or

$$k^2 - 3k - 4 = 0$$

or

$$(k - 4)(k + 1) = 0$$

So we have found the two eigenvalues, $k_1 = 4$ and $k_2 = -1$.

For $k_1 = 4$, the homogeneous system becomes

$$-2x + 3y = 0$$
$$2x - 3y = 0$$

and one solution is $(3, 2)$. Hence, $\alpha_1 = (3, 2)$, or any nonzero multiple of $(3, 2)$, is an eigenvector of T corresponding to the eigenvalue 4.

If $k_2 = -1$, the system
$$3x + 3y = 0$$
$$2x + 2y = 0$$
gives the eigenvector $\alpha_2 = (1, -1)$ corresponding to the eigenvalue $k_2 = -1$.

Relative to the standard basis, the problem $T(\alpha) = k\alpha$ can be written as $AX^T = kX^T$ or as $(A - kI)X^T = 0$ where
$$A = \begin{bmatrix} 2 & 3 \\ 2 & 1 \end{bmatrix}$$

It is easy to check that $AX_1^T = k_1 X_1^T$ and $AX_2^T = k_2 X_2^T$, for
$$\begin{bmatrix} 2 & 3 \\ 2 & 1 \end{bmatrix} \begin{bmatrix} 3 \\ 2 \end{bmatrix} = 4 \begin{bmatrix} 3 \\ 2 \end{bmatrix} \quad \text{and} \quad \begin{bmatrix} 2 & 3 \\ 2 & 1 \end{bmatrix} \begin{bmatrix} 1 \\ -1 \end{bmatrix} = -\begin{bmatrix} 1 \\ -1 \end{bmatrix}$$

In the finite-dimensional case, the eigenvalue problem can always be given as a problem involving matrices. Appropriately we define eigenvalue and eigenvector in terms of matrices.

17-2 Definition

Let A be an n by n matrix, and let X be a nonzero n-tuple such that $AX^T = kX^T$, where k is a scalar. Then X is an *eigenvector* of A corresponding to the *eigenvalue* k. The *characteristic polynomial* of A is $|A - kI|$, a polynomial in k of degree n, and the *characteristic equation* of A is $|A - kI| = 0$.

The eigenvalues are roots of the characteristic equation, because the homogeneous system $(A - kI)X^T = 0$ has a nontrivial solution if and only if $|A - kI| = 0$.

Example 6. Let
$$A = \begin{bmatrix} 5 & -3 & -9 \\ -3 & 5 & 9 \\ 3 & -3 & -7 \end{bmatrix}$$

The characteristic equation is
$$\begin{vmatrix} 5-k & -3 & -9 \\ -3 & 5-k & 9 \\ 3 & -3 & -7-k \end{vmatrix} = 0$$

Lessons in Linear Algebra

or
$$-(k-2)^2(k+1) = 0$$

The characteristic polynomial of A is $-(k-2)^2(k+1)$ or $-k^3 + 3k^2 - 4$. The eigenvalues are $k_1 = k_2 = 2$ and $k_3 = -1$.

With $k_1 = 2$ the system $(A - k_1 I)X^T = 0$ is
$$3x_1 - 3x_2 - 9x_3 = 0$$
$$-3x_1 + 3x_2 + 9x_3 = 0$$
$$3x_1 - 3x_2 - 9x_3 = 0$$

whose solution space is spanned by the two vectors $X_1 = (1, 1, 0)$ and $X_2 = (0, -3, 1)$. Any nontrivial linear combination of X_1 and X_2 is an eigenvector of A corresponding to the eigenvalue 2. (It should be checked that $AX_1^T = 2X_1^T$ and $AX_2^T = 2X_2^T$.)

For the eigenvalue $k_3 = -1$, the system
$$6x_1 - 3x_2 - 9x_3 = 0$$
$$-3x_1 + 6x_2 + 9x_3 = 0$$
$$3x_1 - 3x_2 - 6x_3 = 0$$

gives the eigenvector $X_3 = (1, -1, 1)$. Of course, any nonzero multiple of X_3 is an eigenvector of A corresponding to the eigenvalue -1. Again, you should check to see that $AX_3^T = -X_3^T$.

Next we prove some properties about eigenvalues and eigenvectors. In case $T : V \to V$ where V is of finite dimension, then T can be replaced by a square matrix.

17-3 Theorem   Let k be an eigenvalue of a linear transformation $T : V \to V$. Then the eigenvectors corresponding to k form a subspace of V (called the *eigenspace* of T corresponding to k).

Proof: The eigenspace is simply the kernel of the transformation $T - kI$. ∎

17-4 Theorem   Let T be a linear transformation of V into V, where $\dim V = n$. If k_1 is an eigenvalue of T of multiplicity m, then the corresponding eigenspace has dimension less than or equal to m.

Proof: Let r be the dimension of the corresponding eigenspace and let $\alpha_1, \ldots, \alpha_r$ be linearly independent vectors from the eigenspace. If we represent T by the basis $\{\alpha_1, \ldots, \alpha_r, \ldots, \alpha_n\}$ we get the matrix

$$A = \left\{\begin{array}{c} r \\ \\ \end{array}\right. \left[\begin{array}{cccc|c} k_1 & & & & \\ & k_1 & \bigcirc & & \\ & & \cdot & & \# \\ & \bigcirc & & \cdot & \\ & & & k_1 & \\ \hline & \bigcirc & & & \# \end{array}\right]$$

whose characteristic polynomial has the factor $(k_1 - k)^r$. Hence, $r \leq m$. ∎

If $\alpha_1, \alpha_2, \ldots, \alpha_r$ are eigenvectors corresponding to distinct eigenvalues $k_1, \ldots, k_r$, then $\{\alpha_1, \ldots, \alpha_r\}$ is linearly independent. (The converse does not hold.)

17-5 Theorem

Proof: We are given that $T(\alpha_i) = k_i \alpha_i$ ($i = 1, 2, \ldots, r$), and we plan to show that if $\{\alpha_1, \ldots, \alpha_r\}$ is linearly dependent, then two of the eigenvalues $k_1, \ldots, k_r$ are equal.

Let $\{\alpha_1, \ldots, \alpha_t\}$ be a linearly independent subset of $\{\alpha_1, \ldots, \alpha_r\}$ and suppose that $\alpha_{t+1} = \sum_{i=1}^{t} c_i \alpha_i$, a unique representation of α_{t+1} in terms of $\{\alpha_1, \ldots, \alpha_t\}$, with at least one $c_i \neq 0$. Applying T to α_{t+1} gives the following because of linearity and the fact that each α_i is an eigenvector:

$$k_{t+1} \alpha_{t+1} = \sum_{i=1}^{t} k_i c_i \alpha_i$$

On the other hand, multiplying α_{t+1} by k_{t+1} gives

$$k_{t+1} \alpha_{t+1} = \sum_{i=1}^{t} k_{t+1} c_i \alpha_i$$

Subtracting the two equations, we have

$$\sum_{i=1}^{t} (k_i - k_{t+1}) c_i \alpha_i = 0$$

and since $\{\alpha_1, \ldots, \alpha_t\}$ is linearly independent,

$$(k_i - k_{t+1}) c_i = 0 \quad \text{for} \quad i = 1, \ldots, t$$

Recall that at least one $c_i \neq 0$. Therefore, at least one k_i is equal to k_{t+1}, which means the eigenvalues $k_1, \ldots, k_r$ are not distinct. ∎

Lessons in Linear Algebra

To illustrate Theorem 17-5, look at Example 5. The eigenvalues of $T(x, y) = (2x + 3y, 2x + y)$ are 4 and -1. Their distinctness assures that the corresponding eigenvectors $(3, 2)$ and $(1, -1)$ are linearly independent.

The converse does not hold. That is, suppose the eigenvalues $k_1, \ldots, k_r$ are not distinct. This does not tell us whether or not the corresponding eigenvectors are linearly dependent. Example 6 has two independent eigenvectors corresponding to the multiple eigenvalue 2. The point is further illustrated by Example 7.

Example 7. Let $T(x, y) = (2x, 2y)$ and let $S(x, y) = (2x + y, 2y)$. The transformations have the same eigenvalues, namely $k_1 = k_2 = 2$. But T has two independent eigenvectors $(1, 0)$ and $(0, 1)$, while every eigenvector of S is a multiple of $(1, 0)$.

One of the most profound questions in linear algebra is how to choose bases in order to represent most clearly and usefully a linear transformation. The answer is easy when n eigenvectors of T form a basis of V, because then T can be represented by a "diagonal" matrix, one in which the off-diagonal elements are 0.

17-6 Theorem

Let $T : V \to V$, where dim $V = n$. Then T can be represented by a diagonal matrix if and only if V has a basis of eigenvectors of T.

Proof: Let $\{\alpha_1, \ldots, \alpha_n\}$ be a basis of eigenvectors of T. Then $T(\alpha_i) = k_i \alpha_i$ for $i = 1, \ldots, n$ and the matrix of T is the diagonal matrix

$$\begin{bmatrix} k_1 & & & \\ & k_2 & & \\ & & \ddots & \\ & & & k_n \end{bmatrix}$$

Conversely, if T can be represented by such a diagonal matrix relative to the basis $\{\alpha_1, \ldots, \alpha_n\}$, then $T(\alpha_i) = k_i \alpha_i$ for $i = 1, \ldots, n$. ∎

It is well worth observing that for any diagonal matrix, the elements along the diagonal are eigenvalues of the matrix. This is, in fact, true for "triangular" matrices, matrices which have only zero elements on one side of the main diagonal. For example, the three eigenvalues of

$$\begin{bmatrix} 1 & a & b \\ 0 & 2 & c \\ 0 & 0 & 5 \end{bmatrix}$$

are 1, 2, 5, regardless of the values of a, b, c.

Example 8. We attempt to represent the transformation $T(x, y) = (2x + 3y, 2x + y)$ with a diagonal matrix. In Example 5 we found that $\alpha_1 = (3, 2)$ and $\alpha_2 = (1, -1)$ were eigenvectors corresponding to the eigenvalues 4 and -1, respectively.

Let us represent T relative to the basis $\{\alpha_1, \alpha_2\}$. To do so we must express the images of $\{\alpha_1, \alpha_2\}$ in terms of $\{\alpha_1, \alpha_2\}$.

$$T(\alpha_1) = 4\alpha_1$$
$$T(\alpha_2) = -\alpha_2$$

and so the matrix of T relative to the basis $\{\alpha_1, \alpha_2\}$ is

$$\begin{bmatrix} 4 & 0 \\ 0 & -1 \end{bmatrix}$$

T multiplies vectors parallel to (3, 2) by 2. Vectors parallel to (1, -1) are multiplied by -1.

Example 9. Let $T: R^2 \to R^2$ with $T(x, y) = (-x + 4y, -x + 3y)$. The matrix of T relative to the standard basis is

$$A = \begin{bmatrix} -1 & 4 \\ -1 & 3 \end{bmatrix}$$

The characteristic equation

$$\begin{vmatrix} -1 - k & 4 \\ -1 & 3 - k \end{vmatrix} = 0$$

or

$$k^2 - 2k + 1 = 0$$

gives the repeated eigenvalue $k_1 = 1$ and $k_2 = 1$. The eigenvectors are the solutions of

$$-2x + 4y = 0$$
$$-x + 2y = 0$$

specifically the multiples of $\alpha_1 = (2, 1)$. Since there is no basis of eigenvectors, the transformation T cannot be represented by a diagonal matrix.

Lessons in Linear Algebra

However, we can represent T with a triangular matrix. Simply choose any second vector α_2 such that $\{\alpha_1, \alpha_2\}$ is a basis. Take $\alpha_2 = (0, 1)$. Then

$$T(\alpha_1) = \alpha_1$$
$$T(\alpha_2) = (4, 3) = 2\alpha_1 + \alpha_2$$

and the matrix of T relative to $\{\alpha_1, \alpha_2\}$ is

$$\begin{bmatrix} 1 & 2 \\ 0 & 1 \end{bmatrix}$$

17-7 Corollary (Corollary to Theorem 17-6.) Let $T : V \to V$ where dim $V = n$. If T has n distinct eigenvalues $k_1, \ldots, k_n$, then T can be represented by the diagonal matrix whose diagonal elements are $k_1, \ldots, k_n$.

The proof is left for you. Use Theorems 17-5 and 17-6.

Summary of Topics in Lesson 17

Eigenvalues and eigenvectors of a linear transformation of V into V. Definition and examples.

Eigenvalues and eigenvectors of a matrix. Definitions and a procedure for finding eigenvalues and eigenvectors. The characteristic polynomial of a matrix.

Eigenspace corresponding to a given eigenvalue.

Eigenvectors corresponding to distinct eigenvalues are linearly independent.

Representing a linear transformation $T : V \to V$ by a diagonal matrix is possible if and only if V has a basis consisting of n eigenvectors of T.

If T has n distinct eigenvalues, then T can be represented by a diagonal matrix. The converse does not hold.

If T is represented by a diagonal matrix A, then the diagonal elements of A are eigenvalues of T.

Exercises

1. Find the eigenvalues and eigenvectors of T, given:
   (a) $T(x, y) = (11x - 14y, 7x - 10y)$.

(b) $T(x, y) = (x - y, 5x - y)$. (Here consider T as a mapping of C^2 into itself.)

(c) $T(x, y) = (y, 0)$.

2. Find the eigenvalues and eigenvectors of each of the following matrices. Also try to interpret geometrically the matrices of orders 2 or 3.

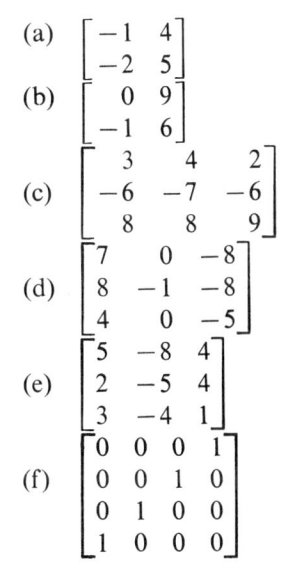

(a) $\begin{bmatrix} -1 & 4 \\ -2 & 5 \end{bmatrix}$

(b) $\begin{bmatrix} 0 & 9 \\ -1 & 6 \end{bmatrix}$

(c) $\begin{bmatrix} 3 & 4 & 2 \\ -6 & -7 & -6 \\ 8 & 8 & 9 \end{bmatrix}$

(d) $\begin{bmatrix} 7 & 0 & -8 \\ 8 & -1 & -8 \\ 4 & 0 & -5 \end{bmatrix}$

(e) $\begin{bmatrix} 5 & -8 & 4 \\ 2 & -5 & 4 \\ 3 & -4 & 1 \end{bmatrix}$

(f) $\begin{bmatrix} 0 & 0 & 0 & 1 \\ 0 & 0 & 1 & 0 \\ 0 & 1 & 0 & 0 \\ 1 & 0 & 0 & 0 \end{bmatrix}$

3. Let D be the differential operator, that is, let $D(f(x)) = f'(x)$.
   (a) Find the eigenvalues and eigenfunctions of D and $2D$.
   (b) Show that for any integer n and constants a, b, $a \sin nx + b \cos nx$ is an eigenfunction of D^2, and find the corresponding eigenvalue.
   (c) Show that e^{ax} (a constant) is an eigenfunction of $D^2 - 7D + 12I$, and find the corresponding eigenvalue.
   (d) Prove that if r is a root of the cubic equation $ax^3 + bx^2 + cx + d = 0$, then e^{rx} is an eigenfunction of $aD^3 + bD^2 + cD + dI$, and find the corresponding eigenvalue.

4. (a) Find the eigenvalues and eigenvectors of the n by n identity matrix I.
   (b) Prove that 0 is an eigenvalue of the square matrix A if and only if A is singular.

5. (a) Show that the eigenvalues of A^T are the same as those of A.
   (b) Let A be invertible. Prove that the eigenvalues of A^{-1} are the

reciprocals of those of A. Also, show that if X is an eigenvector of A corresponding to the eigenvalue k, then X is an eigenvector of A^{-1} corresponding to $1/k$.

(c) Given that X is an eigenvector of A and P is invertible, prove that $P^{-1}X^T$ is an eigenvector of $P^{-1}AP$. How do the two eigenvalues compare?

6. Let T be a linear transformation and let $p(x)$ be a polynomial. Prove: If α is an eigenvector of T corresponding to the eigenvalue k, then α is an eigenvector of $p(T)$ corresponding to the eigenvalue $p(k)$. Illustrate by means of the matrix $A = \begin{bmatrix} 2 & 1 \\ 0 & 2 \end{bmatrix}$ and the eigenvector $X = (1, 0)$, along with the polynomial $p(x) = x^2 + 2x + 1$.

7. Show that $P^{-1}AP$ and A have the same eigenvalues by comparing the determinants of $P^{-1}AP - kI$ and $A - kI$.

8. (a) Explain why any real square matrix of odd order must have a real eigenvalue.

(b) Prove that a real symmetric matrix of order two has two real eigenvalues. (This holds for all real symmetric matrices. In fact, it holds for all Hermitian matrices.)

(c) Let A be a real symmetric matrix. Prove that if X_1 and X_2 are eigenvectors corresponding to distinct eigenvalues k_1 and k_2, then X_1 and X_2 are orthogonal. (Use the definition of eigenvalue-eigenvector and matrix algebra to get $X_2 X_1^T = 0$.)

9. Prove that the sum of the eigenvalues of an n by n matrix A is equal to the trace of A, and that the product of the eigenvalues is equal to $|A|$. Recall that the sum of the roots of the polynomial equation $a_0 x^n + a_1 x^{n-1} + \ldots + a_n = 0$ is $-a_1/a_0$, and the product of the roots is $(-1)^n(a_n/a_0)$.

10. Let T map the space of real functions into itself, with Tf defined by

$$T(f(x)) = f(x + 1)$$

(a) Find a so that $\sin ax$ is an eigenfunction, and find the corresponding eigenvalue.

(b) Show that e^x and e^{-x} are eigenfunctions, and find the corresponding eigenvalues.

(c) Let r be any nonzero real number. Show that r is an eigenvalue and find a corresponding eigenfunction by drawing an analogy with part (b).

Eigenvalues and Eigenvectors

11. Let $T: P_2 \to P_2$ according to the rule $T(p(x)) = p(x - 1)$. Find the eigenvalues and eigenvectors of T.

12. (a) Let $V = C(-\infty, \infty)$ and define $T: V \to V$ by $T(f(x)) = \int_a^x f(t)\, dt$. Show that T has no eigenvalue.

    (b) Let V be the set of functions f for which $\int_{-\infty}^x f(t)\, dt$ exists for every real number x. Find the eigenvalues and eigenfunctions of T, given that $T(f(x)) = \int_{-\infty}^x f(t)\, dt$.

13. Let V be the space consisting of all convergent sequences of real numbers. The typical element of V is $\alpha = \{x_n\} = \{x_1, x_2, \ldots\}$, where x_n converges to some number x. Define $T: V \to V$ by $T\{x_n\} = \{x - x_n\}$. That is, $T(x_1, x_2, \ldots) = (x - x_1, x - x_2, \ldots)$. After checking that T is linear, find the eigenvalues and eigenvectors of T.

14. Let $T: R^2 \to R^2$ by $T(x, y) = (x + 2y, 2x + y)$. Find a diagonal matrix which represents T.

15. Find a diagonal matrix which represents the transformation
$$T(x, y, z, w) = (w, z, y, x)$$
(See Exercise 2(f).)

Remarks on Exercises

1. (a) $k_1 = 4$, $\alpha_1 = (2, 1)$. $k_2 = -3$, $\alpha_2 = (1, 1)$. (Of course, any multiple of an eigenvector is also an eigenvector.)
   (b) $k_1 = 2i$, $\alpha_1 = (1, 1 - 2i)$. $k_2 = -2i$, $\alpha_2 = (1, 1 + 2i)$.
   (c) $k_1 = k_2 = 0$. $(1, 0)$ is an eigenvector.

2. (a) $k_1 = 1$, $\alpha_1 = (2, 1)$. $k_2 = 3$, $\alpha_2 = (1, 1)$.
   (b) $k_1 = k_2 = 3$. $(3, 1)$ is an eigenvector.
   (c) $k_1 = 1$, $\alpha_1 = (1, 0, -1)$. $k_2 = 5$, $\alpha_2 = (0, 1, -2)$. $k_3 = -1$, $\alpha_3 = (1, -1, 0)$.
   (d) $k_1 = 3$, $\alpha_1 = (2, 2, 1)$. $k_2 = k_3 = -1$, $\alpha_2 = (0, 1, 0)$ and $\alpha_3 = (1, 0, 1)$ are a basis for the eigenspace corresponding to the eigenvalue -1.
   (e) $k_1 = 3$, $\alpha_1 = (2, 1, 1)$. $k_2 = k_3 = -1$, corresponding eigenvector $(2, 2, 1)$.
   (f) $k_1 = k_2 = 1$. A basis for the eigenspace is $\{(1, 0, 0, 1), (0, 1, 1, 0)\}$. $k_3 = k_4 = -1$. A basis for the eigenspace is $(1, 0, 0, -1), (0, 1, -1, 0)$.

211

Lessons in Linear Algebra

3. (a) Every real number k is an eigenvalue of D, with e^{kx} (or any nonzero multiple of e^{kx}) as a corresponding eigenfunction. Every real number k is an eigenvalue of $2D$, with corresponding eigenfunction $e^{kx/2}$.
   (b) The eigenvalue is $-n^2$.
   (c) The eigenvalue is $a^2 - 7a + 12$.
   (d) The eigenvalue is $ar^3 + br^2 + cr + d = 0$.

4. (a) The only eigenvalue is 1. Every vector is an eigenvector.
   (b) $|A - 0I| = 0$ if and only if $|A| = 0$, or if and only if A is singular.

5. (a) Use the fact that the det $(M^T) = $ det M. Then $|A^T - kI| = |A - kI|$, and the matrices A and A^T have the same characteristic polynomials.
   (b) Since A is invertible, let k be a nonzero eigenvalue of A, corresponding to the eigenvector X. Then $AX^T = kX^T$, $X^T = kA^{-1}X^T$, and $A^{-1}X^T = (1/k)X^T$. Hence, X is an eigenvector of A^{-1} corresponding to the eigenvalue $1/k$.
   (c) Given $AX^T = kX^T$, consider $P^{-1}APP^{-1}X^T = kP^{-1}X^T$. The eigenvalues are the same.

6. In the illustrative example $AX^T = 2X^T$, so the eigenvalue is 2. The matrix $p(A) = A^2 + 2A + I$ has eigenvalue $p(2) = 9$.

7. $|P^{-1}AP - kI| = |P^{-1}(A - kI)P| = |P^{-1}| |A - kI| |P| = |A - kI|$.

8. (a) Every real polynomial equation of odd order has a real root.
   (b) Look at the discriminant of the characteristic equation of $\begin{bmatrix} a & b \\ b & c \end{bmatrix}$.
   (c) Since $AX_1^T = k_1X_1^T$, we have $X_2AX_1^T = k_1X_2X_1^T$. Similarly $X_1AX_2^T = k_2X_1X_2^T$. Transposing the last equation gives $X_2AX_1^T = k_2X_2X_1^T$, since $A^T = A$. Therefore, $k_1X_2X_1^T = k_2X_2X_1^T$, $(k_1 - k_2)X_2X_1^T = 0$, and $X_2X_1^T = 0$.

9. Let $f(k) = a_0k^n + a_1k^{n-1} + \ldots + a_n$ be the characteristic polynomial of A. The sum of the eigenvalues of A is $-a_1/a_0$ and the product of the eigenvalues is $(-1)^n a_n/a_0$. Inspecting the determinant of $A - kI$ shows that $a_0 = (-1)^n$ and $a_1 = (-1)^{n-1}Tr(A)$. Setting $k = 0$ in the determinant of $A - kI$ shows that $|A| = a_n$. Therefore, if $k_1, \ldots, k_n$ are the eigenvalues, then $\sum k_i = Tr(A)$ and the product of the eigenvalues is $\Pi k_i = |A|$.

10. (a) Eigenvalues 1, -1. Corresponding eigenfunctions $\sin 2\pi x$, $\sin \pi x$.

 (b) The eigenvalues are e and e^{-1}.
 (c) r^x is an eigenfunction

11. The eigenvalue is 1, the eigenvectors are the constants.

12. (a) If k were an eigenvalue then an eigenfunction would satisfy $\int_a^x f(t)\,dt = kf(x)$. Differentiating gives $f(x) = kf'(x)$, implying that $f(x) = ce^{x/k}$. But $T(e^{x/k}) \neq ke^{x/k}$, so $e^{x/k}$ is not an eigenfunction.

 (b) Observe that $T(f(x)) = \int_{-\infty}^0 f(t)\,dt + \int_0^x f(t)\,dt$. If k were an eigenvalue, then differentiation shows that the eigenfunction satisfies $f(x) = kf'(x)$. Therefore, $e^{x/k}$ is an eigenfunction. We conclude that every number k is an eigenvalue with $ce^{x/k}$ as a corresponding eigenfunction, provided $k \neq 0$.

13. If $\{x_n\}$ converges to 0, then it is an eigenfunction corresponding to the eigenvalue -1. If $\{x_n\}$ is a constant sequence $\{a, a, a, \ldots\}$, $a \neq 0$, then it is an eigenfunction corresponding to the eigenvalue 0.

14. There is a basis of eigenvectors, $\{(1, 1), (1, -1)\}$. Relative to this basis, the matrix of T is $\begin{bmatrix} 3 & 0 \\ 0 & -1 \end{bmatrix}$.

15. Relative to the basis $\{(1, 0, 0, 1), (0, 1, 1, 0), (1, 0, 0, -1), (0, 1, -1, 0)\}$, the matrix of T is
$$\begin{bmatrix} 1 & 0 & 0 & 0 \\ 0 & 1 & 0 & 0 \\ 0 & 0 & -1 & 0 \\ 0 & 0 & 0 & -1 \end{bmatrix}$$

18

SIMILARITY AND CANONICAL FORMS

In this lesson we consider how to select a basis which best represents a linear transformation T of a finite-dimensional space into itself. T can always be represented by a triangular matrix and under some circumstances T can be represented by a diagonal matrix.

Matrices which represent the same linear transformation, called similar matrices, are of great interest. We will show how to decide whether or not a given matrix A is similar to a diagonal matrix and how to construct a matrix P so that $P^{-1}AP$ is diagonal or triangular.

Also, we will discuss some of the properties and applications of similar matrices.

An equivalence relation on a set S partitions S into disjoint subsets, and conversely any such partition of S defines an equivalence relation. If S is partitioned into the disjoint subsets $\{T_i\}$, then it is sometimes convenient to represent each of the subsets with one of its elements, This is something you have already done. A rational number is of the form a/b, where a and b are integers and $b \neq 0$. An equivalence class is $\{1/2, 2/4, -30/60, \ldots\}$, and it is common to select $1/2$ as the representative of this equivalence class.

We will take the same approach in classifying the linear transformations of an n-dimensional linear space V into V. Or what amounts to the same thing, we will classify the set of n by n matrices and from each equivalence class we will select a representative. This involves selecting an appropriate basis and knowing how to change the bases which are used to produce matrices representing T.

First we consider the problem of changing bases.

18-1 Definition Let $\mathcal{A} = \{\alpha_1, \ldots, \alpha_n\}$ and $\mathcal{B} = \{\beta_1, \ldots, \beta_n\}$ be bases of V. The *matrix of transition from $\mathcal{A}$ to $\mathcal{B}$* is $p = [p_{ij}]_{n,n}$, where p_{ij} is defined by

$$\beta_j = \sum_i p_{ij}\alpha_i, \quad j = 1, \ldots, n$$

Similarity and Canonical Forms

18-2 Theorem

Let P be the transition matrix from the basis $\mathcal{A} = \{\alpha_1, \ldots, \alpha_n\}$ to the basis $\mathcal{B} = \{\beta_1, \ldots, \beta_n\}$, and let $\alpha \in V$. Let $X = (x_1, \ldots, x_n)$ and $X' = (x'_1, \ldots, x'_n)$ be the coordinates of α relative to $\mathcal{A}$ and $\mathcal{B}$, respectively. Then $(X')^T = P^{-1}X^T$ or $X^T = P(X')^T$.

Proof: We are given $\alpha = \sum_i x_i \alpha_i = \sum_j x'_j \beta_j$. Hence,

$$\sum_i x_i \alpha_i = \sum_j x'_j \sum_i p_{ij} \alpha_i = \sum_i \left(\sum_j p_{ij} x'_j \right) \alpha_i$$

and

$$x_i = \sum_j p_{ij} x'_j$$

This gives $X^T = P(X')^T$. ∎

Example 1. In R^2, let $T(x, y) = (3x - 2y, 2x - y)$, and let $\mathcal{A} = \{(1, 1), (1, -1)\}$, $\mathcal{B} = \{(3, 1), (1, -3)\}$ be two bases. The matrix of transition P is found by expressing the basis $\mathcal{B} = \{\beta_1, \beta_2\}$ in terms of $\mathcal{A} = \{\alpha_1, \alpha_2\}$:

$$\beta_1 = 2\alpha_1 + \alpha_2$$
$$\beta_2 = -\alpha_1 + 2\alpha_2$$

and

$$P = \begin{vmatrix} 2 & -1 \\ 1 & 2 \end{vmatrix}$$

Now let α be any vector of R^2 and let $X = (x_1, x_2)$, $X' = (x'_1, x'_2)$ be the coordinates of α relative to $\mathcal{A}$ and $\mathcal{B}$, respectively. Theorem 18-2 asserts that $X^T = P(X')^T$.

We illustrate (Figure 18-1) with the vector $\alpha = (5, 1)$. Then $\alpha = 3\alpha_1 + 2\alpha_2 = (8/5)\beta_1 + (1/5)\beta_2$. Then

$$X^T = P(X')^T$$

because

$$\begin{bmatrix} 3 \\ 2 \end{bmatrix} = \begin{bmatrix} 2 & -1 \\ 1 & 2 \end{bmatrix} \begin{bmatrix} \frac{8}{5} \\ \frac{1}{5} \end{bmatrix}$$

$$X = (3, 2) \qquad X' = \left(\frac{8}{5}, \frac{1}{5}\right) \qquad P = \begin{bmatrix} 2 & -1 \\ 1 & 2 \end{bmatrix}$$

Lessons in Linear Algebra

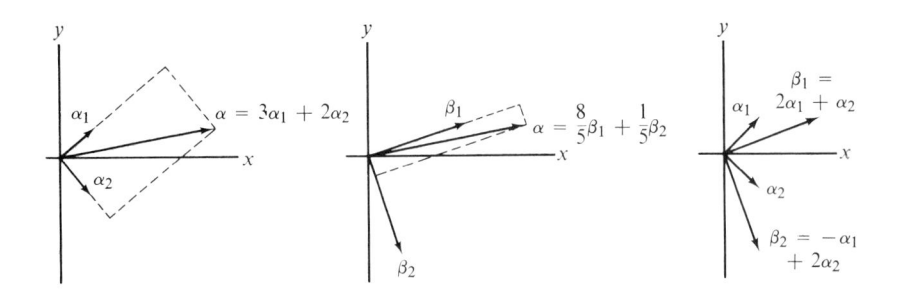

FIGURE 18-1

So far we have said nothing about T. You will recall that, relative to any basis, T is represented by a matrix. Let A and B be the matrices representing T relative to the bases $\mathcal{A}$, $\mathcal{B}$, respectively. To find A we express $T(\alpha_1)$, $T(\alpha_2)$ in terms of α_1, α_2. The rule $T(x, y) = (3x - 2y, 2x - y)$ gives

$$T(\alpha_1) = (1, 1) = \alpha_1$$
$$T(\alpha_2) = (5, 3) = 4\alpha_1 + \alpha_2$$

and so

$$A = \begin{bmatrix} 1 & 4 \\ 0 & 1 \end{bmatrix}$$

Similarly,

$$T(\beta_1) = (7, 5) = \frac{13}{5}\beta_1 - \frac{4}{5}\beta_2$$
$$T(\beta_2) = (9, 5) = \frac{16}{5}\beta_1 - \frac{3}{5}\beta_2$$
$$B = \frac{1}{5}\begin{bmatrix} 13 & 16 \\ -4 & -3 \end{bmatrix}$$

There is an important relation between B and A. It turns out that $P^{-1}AP = B$, a fact that we generalize.

18-3 Theorem

Let A and B be n by n matrices representing a linear transformation $T: V \to V$ relative to the bases $\mathcal{A}$ and $\mathcal{B}$, respectively, and let P be the matrix of transition from $\mathcal{A}$ to $\mathcal{B}$. Then $B = P^{-1}AP$.

Proof: The proof is not difficult. It is just a question of placing things in the proper order. Suppose $\alpha \in V$, and let:

Similarity and Canonical Forms

$X = (x_1, \ldots, x_n)$ be the coordinate of α relative to $\mathcal{A}$.
$X' = (x'_1, \ldots, x'_n)$ the coordinate of α relative to $\mathcal{B}$.
$Y = (y_1, \ldots, y_n)$ the coordinate of $T(\alpha)$ relative to $\mathcal{A}$.
$Y' = (y'_1, \ldots, y'_n)$ the coordinate of $T(\alpha)$ relative to $\mathcal{B}$.

Then we have

$Y^T = AX^T$ (A is the matrix of T relative to $\mathcal{A}$.)
$P^{-1}Y^T = P^{-1}AX^T$
$(Y')^T = P^{-1}AP(X')^T$ (By Theorem 18-3, $(Y')^T = P^{-1}Y^T$ and $X^T = P(X')^T$.)

Also

$(Y')^T = B(X')^T$ (B is the matrix of T relative to $\mathcal{B}$.)

Therefore,

$(P^{-1}AP)(X')^T = B(X')^T$ for all X' and $P^{-1}AP = B$ ∎

18-4 Definition

The matrix A is *similar* to the matrix B if there exists a matrix P such that $P^{-1}AP = B$.

Similarity is an equivalence relation because: (1) A is similar to itself since $I^{-1}AI = A$; (2) if A is similar to B by means of $P^{-1}AP = B$, then B is similar to A by $(P^{-1})^{-1}BP^{-1} = A$; (3) if A is similar to B, and B to C, then A is similar to C.

In Lesson 17 we showed that a linear transformation T of an n-dimensional space into itself can be represented by a diagonal matrix if and only if T has n linearly independent eigenvectors. Given any matrix A representing such a transformation, there is a straightforward way of finding P so that $P^{-1}AP$ is diagonal.

18-5 Theorem

Let A be an n by n matrix with n linearly independent eigenvectors $X_1, \ldots, X_n$ and let $P = [X_1^T, X_2^T, \ldots, X_n^T]$, the matrix which has for its columns the eigenvectors of T. Then

$$P^{-1}AP = D = \begin{bmatrix} k_1 & & & \\ & k_2 & & \\ & & \ddots & \\ & & & k_n \end{bmatrix}$$

where k_i is the eigenvalue corresponding to X_i.

Lessons in Linear Algebra

Proof: We will show $AP = PD$.

$$\begin{align}
AP &= A[X_1^T, X_2^T, \ldots, X_n^T] \\
&= [AX_1^T, AX_2^T, \ldots, AX_n^T] && \text{(matrix multiplication)} \\
&= [k_1 X_1^T, k_2 X_2^T, \ldots, k_n X_n^T] && (AX_i^T = k_i X_i^T \text{ because of eigen-} \\
& && \text{value-eigenvector relation.)} \\
&= [X_1^T, X_2^T, \ldots, X_n^T] D && \text{(matrix multiplication)} \\
&= PD
\end{align}$$

Therefore, $P^{-1}AP = D$. ∎

Example 2. Let

$$A = \begin{bmatrix} 1 & 6 \\ 5 & 2 \end{bmatrix}$$

Then $X_1 = (1, 1)$ and $X_2 = (6, -5)$ are eigenvectors with corresponding eigenvalues $k_1 = 7$, $k_2 = -4$. Construct

$$P = \begin{bmatrix} 1 & 6 \\ 1 & -5 \end{bmatrix}$$

Then

$$P^{-1}AP = \frac{1}{11}\begin{bmatrix} 5 & 6 \\ 1 & -1 \end{bmatrix}\begin{bmatrix} 1 & 6 \\ 5 & 2 \end{bmatrix}\begin{bmatrix} 1 & 6 \\ 1 & -5 \end{bmatrix} = \begin{bmatrix} 7 & 0 \\ 0 & -4 \end{bmatrix} = D$$

Example 3. We use Example 2 to solve the system of differential equations

$$\frac{dy_1}{dx} = y_1 + 6y_2$$

$$\frac{dy_2}{dx} = 5y_1 + 2y_2$$

We let $Y = [y_1 \ y_2]$, $A = \begin{bmatrix} 1 & 6 \\ 5 & 2 \end{bmatrix}$ and write

$$\frac{dY^T}{dx} = AY^T$$

A substitution $Y^T = PZ^T$ with P nonsingular gives

$$P\frac{dZ^T}{dx} = APZ^T$$

or

$$\frac{dZ^T}{dx} = (P^{-1}AP)Z^T$$

We choose $P = \begin{bmatrix} 1 & 6 \\ 1 & -5 \end{bmatrix}$, from Example 2. Then $dZ^T/dx = DZ^T$. where $D = \begin{bmatrix} 7 & 0 \\ 0 & -4 \end{bmatrix}$. Then $dz_1/dx = 7z_1$, $dz_2/dx = -4z_2$, and $z_1 = c_1e^{7x}$, $z_2 = c_2e^{-4x}$.

The connection $Y^T = PZ^T$ gives

$$y_1 = c_1e^{7x} + 6c_2e^{-4x}$$
$$y_2 = c_1e^{7x} - 5c_2e^{-4x}$$

Not every matrix is similar to a diagonal matrix, but every matrix is similar to a triangular matrix. In order to discuss this further, we develop the multiplication of partitioned matrices.

Example 4. Let

$$A = \begin{vmatrix} 1 & 3 & 2 & -1 \\ 0 & 1 & 0 & 2 \\ \hline -1 & 3 & 4 & 2 \end{vmatrix}, \quad B = \begin{vmatrix} 1 & 0 & 1 & 0 \\ 0 & 1 & 0 & -1 \\ 0 & 0 & 1 & 1 \\ \hline 1 & 1 & 1 & 1 \end{vmatrix}$$

be two matrices. A is of type (3, 4), B of type (4, 4), so the product AB is of type (3, 4). We partition A and B into submatrices by using the dotted lines, and we have

$$A = \begin{vmatrix} A_{11} & A_{12} \\ A_{21} & A_{22} \end{vmatrix}, \quad B = \begin{vmatrix} B_{11} & B_{12} \\ B_{21} & B_{22} \end{vmatrix}$$

where

$$A_{11} = \begin{bmatrix} 1 & 3 \\ 0 & 1 \end{bmatrix}, \quad A_{12} = \begin{bmatrix} 2 & -1 \\ 0 & 2 \end{bmatrix}, \quad A_{21} = [-1 \ \ 3],$$

$$A_{22} = [4 \ \ 2], \quad B_{11} = \begin{bmatrix} 1 & 0 & 1 \\ 0 & 1 & 0 \end{bmatrix}, \quad B_{12} = \begin{bmatrix} 0 \\ -1 \end{bmatrix},$$

$$B_{21} = \begin{bmatrix} 0 & 0 & 1 \\ 1 & 1 & 1 \end{bmatrix}, \quad B_{22} = \begin{bmatrix} 1 \\ 1 \end{bmatrix}$$

Multiplying A and B as partitioned matrices gives

Lessons in Linear Algebra

$$= \begin{bmatrix} A_{11}B_{11} + A_{12}B_{21} & A_{11}B_{12} + A_{12}B_{22} \\ A_{21}B_{11} + A_{22}B_{21} & A_{21}B_{12} + A_{22}B_{22} \end{bmatrix}$$

$$= \begin{bmatrix} \begin{bmatrix} 1 & 3 & 1 \\ 0 & 1 & 0 \end{bmatrix} + \begin{bmatrix} -1 & -1 & 1 \\ 2 & 2 & 2 \end{bmatrix} & \begin{bmatrix} -3 \\ -1 \end{bmatrix} + \begin{bmatrix} 1 \\ 2 \end{bmatrix} \\ [\,-1 \quad 3 \quad -1\,] + [\,2 \quad 2 \quad 6\,] & [-3] + [6] \end{bmatrix}$$

$$= \begin{bmatrix} \begin{bmatrix} 0 & 2 & 2 \\ 2 & 3 & 2 \end{bmatrix} & \begin{bmatrix} -2 \\ 1 \end{bmatrix} \\ [\,1 \quad 5 \quad 5\,] & 3 \end{bmatrix}$$

Considered as a 3 by 4 matrix, the preceding is AB.

The result is general. Furthermore, any partitioning of A and B into submatrices $[A_{ij}]$ and $[B_{ij}]$ is allowed just so long as $[A_{ij}]$ and $[B_{ij}]$ are conformable for multiplication.

18-6 Theorem

Let $A = [a_{ij}]_{m,n}$, $B = [b_{ij}]_{n,r}$ be partitioned into submatrices A', B', with
$$A' = [A_{ij}]_{p,q} \qquad B' = [B_{ij}]_{q,s}$$
Then (except for the bracketing), $A'B' = AB$.

Proof: This requires only a comparison of the ij-elements of AB and $A'B'$. It is not essential that one write out a formal proof. ∎

Next we show that every square matrix is similar to a "triangular" matrix, a matrix with all zeros on one side of the main diagonal. The proof will be constructive in that it gives a method for reducing a matrix to a triangular form.

18-7 Theorem

Let A be an n by n matrix. Then A is similar to an "upper triangular" matrix B, where $b_{ij} = 0$ when $i > j$.

Proof: We use mathematical induction. For $n = 1$, the theorem is trivial. Assume the theorem holds for all matrices of order less than n, and let P be any nonsingular matrix whose first column is an eigenvector of A. That is, let $P = [C_1, C_2, \ldots, C_n]$ where $AC_1 = k_1 C_1$, and P is nonsingular. Then the first column of AP is $k_1 C_1$, or

Similarity and Canonical Forms

$$P\begin{bmatrix} k_1 \\ 0 \\ \cdot \\ \cdot \\ \cdot \\ 0 \end{bmatrix}$$

This means that the first column of $P^{-1}AP$ is

$$\begin{bmatrix} k_1 \\ 0 \\ \cdot \\ \cdot \\ \cdot \\ 0 \end{bmatrix}$$

and hence

$$P^{-1}AP = \begin{bmatrix} k_1 & B_{12} \\ 0 & B_{22} \end{bmatrix}$$

where 0 is of type $(n-1, 1)$, B_{12} of type $(1, n-1)$, and B_{22} of type $(n-1, n-1)$.

By the induction hypothesis there is a matrix Q such that $Q^{-1}B_{22}Q$ is (upper) triangular. Let

$$R = \begin{bmatrix} 1 & 0 \\ 0 & Q \end{bmatrix}$$

and we have

$$(PR)^{-1}A(PR) = R^{-1}\begin{bmatrix} k_1 & B_{12} \\ 0 & B_{22} \end{bmatrix} R$$
$$= \begin{bmatrix} 1 & 0 \\ 0 & Q^{-1} \end{bmatrix}\begin{bmatrix} k_1 & B_{12} \\ 0 & B_{22} \end{bmatrix}\begin{bmatrix} 1 & 0 \\ 0 & Q \end{bmatrix}$$
$$= \begin{bmatrix} k_1 & B_{12}Q \\ 0 & Q^{-1}B_{22}Q \end{bmatrix}$$

a triangular matrix.

Example 5. Let

$$A = \begin{bmatrix} 0 & 0 & 1 \\ -2 & 0 & 3 \\ -2 & -1 & 4 \end{bmatrix}$$

Lessons in Linear Algebra

The characteristic polynomial of A is $-(k - 1)^2(k - 2)$ and so the eigenvalues are 1, 1, 2. Corresponding to the eigenvalue 1 is the eigenvector (1, 1, 1). Using the eigenvector as the first column, we construct the nonsingular matrix

$$P = \begin{bmatrix} 1 & 0 & 0 \\ 1 & 1 & 0 \\ 1 & 0 & 1 \end{bmatrix}$$

Then

$$P^{-1} = \begin{bmatrix} 1 & 0 & 0 \\ -1 & 1 & 0 \\ -1 & 0 & 1 \end{bmatrix} \quad \text{and} \quad P^{-1}AP = \begin{bmatrix} 1 & 0 & 1 \\ 0 & 0 & 2 \\ 0 & -1 & 3 \end{bmatrix}$$

Now we work with

$$B_{22} = \begin{bmatrix} 0 & 2 \\ -1 & 3 \end{bmatrix}$$

Its characteristic polynomial is $(k - 1)(k - 2)$, a factor of the characteristic polynomial of A. The eigenvalues of B_{22} are 1 and 2, and corresponding eigenvectors are (2, 1) and (1, 1). Let

$$Q = \begin{bmatrix} 2 & 1 \\ 1 & 1 \end{bmatrix}$$

Then

$$Q^{-1}B_{22}Q = \begin{bmatrix} 1 & 0 \\ 0 & 2 \end{bmatrix},$$

a triangular (in fact, diagonal) matrix.

Take

$$R = \begin{bmatrix} 1 & 0 \\ 0 & Q \end{bmatrix} = \begin{bmatrix} 1 & 0 & 0 \\ 0 & 2 & 1 \\ 0 & 1 & 1 \end{bmatrix}$$

Then

$$PR = \begin{bmatrix} 1 & 0 & 0 \\ 1 & 2 & 1 \\ 1 & 1 & 1 \end{bmatrix}$$

and

$$(PR)^{-1}A(PR) = R^{-1} \begin{bmatrix} 1 & 0 & 1 \\ 0 & 0 & 2 \\ 0 & -1 & 3 \end{bmatrix} R$$

Similarity and Canonical Forms

$$= \begin{bmatrix} 1 & 0 & 0 \\ 0 & 1 & -1 \\ 0 & -1 & 2 \end{bmatrix} \begin{bmatrix} 1 & 0 & 1 \\ 0 & 0 & 2 \\ 0 & -1 & 3 \end{bmatrix} \begin{bmatrix} 1 & 0 & 0 \\ 0 & 2 & 1 \\ 0 & 1 & 1 \end{bmatrix}$$

$$= \begin{bmatrix} 1 & 1 & 1 \\ 0 & 1 & 0 \\ 0 & 0 & 2 \end{bmatrix}$$

The matrix A is not similar to a diagonal matrix because A does not have n linearly independent eigenvectors.

Changing a matrix A into a similar matrix $P^{-1}AP = B$ preserves a number of properties of A.

18-8 Theorem

Let A and B be similar matrices, with $P^{-1}AP = B$. Then

(a) A and B have the same characteristic polynomial, and hence the same eigenvalues.
(b) $\det A = \det B$.
(c) if $f(x)$ is any polynomial, then $f(B) = f(P^{-1}AP) = P^{-1}f(A)P$.

Proof: (a) We start with the characteristic polynomial of B.

$$|P^{-1}AP - kI| = |P^{-1}AP - kP^{-1}P|$$
$$= |P^{-1}(A - kI)P| \quad \text{(distributive laws)}$$
$$= |P^{-1}| \, |A - kI| \, |P| \quad \text{(property of determinants)}$$
$$= |A - kI| \quad (|P^{-1}P| = 1)$$

Parts (b) and (c) are left for the student. To handle (c) note that $(P^{-1}AP)^n = P^{-1}A^nP$ and $P^{-1}(M + N)P = P^{-1}MP + P^{-1}NP$. ∎

Example 6. It is possible that a real matrix has imaginary eigenvalues and therefore cannot be similar to a diagonal or triangular real matrix. Consider

$$A = \begin{bmatrix} 1 & 2 \\ -1 & -1 \end{bmatrix}$$

The characteristic polynomial is $k^2 + 1$, so there is no real matrix P such that $P^{-1}AP$ is diagonal or triangular, because the diagonal elements would have to be eigenvalues of $P^{-1}AP$ and of A.

However, if we use the eigenvectors $(2, -1 + i)$ and $(2, -1 - i)$ to construct

Lessons in Linear Algebra

$$P = \begin{bmatrix} 2 & 2 \\ -1+i & -1-i \end{bmatrix}$$

then

$$P^{-1} = \frac{1}{4}\begin{bmatrix} 1-i & -2i \\ 1+i & 2i \end{bmatrix}$$

and

$$P^{-1}AP = \begin{bmatrix} i & 0 \\ 0 & -i \end{bmatrix}$$

the expected diagonal matrix.

The lesson is concluded with the celebrated Hamilton-Cayley theorem.*

18-9 Theorem

Every square matrix satisfies its own characteristic equation.

Proof: Choose P so that $P^{-1}AP = B$ is triangular. Then

$$B = \begin{bmatrix} k_1 & & & \# \\ & k_2 & & \\ & & \ddots & \\ & \bigcirc & & k_n \end{bmatrix}$$

where $k_1, \ldots, k_n$ are (complex) eigenvalues of A. Also let $f(x)$ be the characteristic polynomial of A and of B. Then $f(x) = (k_1 - x)(k_2 - x) \ldots (k_n - x)$.

We propose to show that $f(B) = 0$.

$$f(B) = (k_1 I - B)(k_2 I - B) \ldots (k_n I - B)$$

or

---

*Arthur Cayley (1821–1895), one of England's greatest mathematicians, is noted for his work in n-dimensional geometry, algebraic systems, and theory of matrices. The great Irish mathematician William Rowan Hamilton (1805–1865) is remembered for his work in applying mathematics to physical sciences such as optics and mechanics. He is especially remembered for his work in vectors and for his creation of "quarternions," a generalization of complex numbers.

Cayley and Hamilton worked at a time when the state of mathematics in the British Isles had lagged behind that on the continent.

Similarity and Canonical Forms

$$f(B) = \begin{bmatrix} 0 & \# & \# & \# \\ & \# & \# & \\ & & \ddots & \\ & \bigcirc & & \ddots \\ & & & \# \end{bmatrix} \begin{bmatrix} \# & \# & \# \\ 0 & \# & \\ & \ddots & \\ \bigcirc & & \ddots \\ & & \# \end{bmatrix} \cdots$$

$$\begin{bmatrix} \# & \# & \# \\ & \# & \\ & & \ddots \\ \bigcirc & & \ddots \\ & & \# \\ & & & 0 \end{bmatrix}$$

the product of n triangular matrices, such that the ii-element of the ith matrix is 0. Multiplication shows the following:

The first two columns of $(k_1I - B)(k_2I - B)$ are zero.
The first three columns of $(k_1I - B)(k_2I - B)(k_3I - B)$ are zero.
.
.
.

All the columns of $f(B)$ are 0.

So we have $f(B) = 0$, but we need to show $f(A) = 0$. Since $B = P^{-1}AP$, by Theorem 18-8(c) we have

$$0 = f(B) = f(P^{-1}AP) = P^{-1}f(A)P$$

and therefore $f(A) = 0$. ∎

The proof is valid for all complex matrices (including real matrices) even though some of the eigenvalues may be imaginary.

Example 6. Let

$$A = \begin{bmatrix} 1 & 3 \\ 4 & 2 \end{bmatrix}$$

The characteristic polynomial is

$$f(x) = \begin{bmatrix} 1-x & 3 \\ 4 & 2-x \end{bmatrix} = x^2 - 3x - 10$$

Lessons in Linear Algebra

Then
$$f(A) = A^2 - 3A - 10I = \begin{bmatrix} 13 & 9 \\ 12 & 16 \end{bmatrix} - \begin{bmatrix} 3 & 9 \\ 12 & 6 \end{bmatrix} - \begin{bmatrix} 10 & 0 \\ 0 & 10 \end{bmatrix} = 0$$

We may also note that since $A(A - 3I) = 10I$,

$$A^{-1} = \frac{1}{10}(A - 3I) = \frac{1}{10}\begin{bmatrix} -2 & 3 \\ 4 & -1 \end{bmatrix}$$

Example 7. Let

$$B = \begin{bmatrix} 2 & 1 & 3 \\ 0 & i & -1 \\ 0 & 0 & -i \end{bmatrix}$$

The characteristic polynomial
$$f(x) = (2 - x)(i - x)(-i - x)$$

or
$$f(x) = -x^3 + 2x^2 - x + 2$$

In order to check the Hamilton-Cayley theorem, we may verify that $-B^3 + 2B^2 - B + 2I = 0$. Instead, we compute

$$f(B) = (2I - B)(iI - B)(-iI - B)$$
$$= \begin{bmatrix} 0 & -1 & -3 \\ 0 & 2-i & 1 \\ 0 & 0 & 2+i \end{bmatrix} \begin{bmatrix} -2+i & -1 & -3 \\ 0 & 0 & 1 \\ 0 & 0 & 2i \end{bmatrix} \begin{bmatrix} -2-i & -1 & -3 \\ 0 & -2i & 1 \\ 0 & 0 & 0 \end{bmatrix}$$
$$= \begin{bmatrix} 0 & 0 & -1-6i \\ 0 & 0 & 2+i \\ 0 & 0 & -2+4i \end{bmatrix} \begin{bmatrix} -2-i & -1 & -3 \\ 0 & -2i & 1 \\ 0 & 0 & 0 \end{bmatrix}$$
$$= 0$$

We have observed that any square matrix is similar to a triangular matrix and that an n by n matrix with n independent eigenvectors is similar to a diagonal matrix. Actually we can do better than the triangular form for a general matrix. Any square matrix is similar to a matrix with all zeros except in the main diagonal and the "super" diagonal, the line just above the main diagonal.

Few details will be discussed here, but a typical "Jordan normal form" is

Similarity and Canonical Forms

$$\begin{bmatrix} 2 & 1 & 0 & 0 & 0 \\ 0 & 2 & 0 & 0 & 0 \\ 0 & 0 & 3 & 1 & 0 \\ 0 & 0 & 0 & 3 & 1 \\ 0 & 0 & 0 & 0 & 3 \end{bmatrix}$$

Every element in the super diagonal must be 1 or 0. In case the matrix has n independent eigenvectors then the Jordan normal form is diagonal. (For further details see Evar D. Nering, *Linear Algebra and Matrix Theory*, John Wiley and Sons, New York, 1963 and Paul Halmos, *Finite Dimensional Vector Spaces*, D. Van Nostrand, Princeton, New Jersey, 1958.)

Summary of Topics in Lesson 18

Change of basis. The matrix P of transition from one basis to another relates new and old coordinates by $(X')^T = (P^{-1})X^T$.

The matrix of T relative to the new basis is $P^{-1}AP$, where A is the matrix of T relative to the old basis, and P is the transition matrix.

Similarity of matrices.

The $n \times n$ matrix A is similar to a diagonal matrix if and only if A has n independent eigenvectors.

How to construct P so that $P^{-1}AP$ is diagonal or triangular. Similar matrices have the same determinant and the same characteristic polynomial.

Any square matrix satisfies its characteristic equation (Hamilton-Cayley).

Exercises

1. Let

$$A = \begin{bmatrix} 1 & -5 \\ 1 & -1 \end{bmatrix}$$

   Show that as a real matrix A is not similar to a diagonal matrix, but as a complex matrix A is similar to a diagonal matrix. Also display P so that $P^{-1}AP$ is diagonal.

2. (a) Complete the proof that similarity is an equivalence relation by showing that it is transitive.

Lessons in Linear Algebra

    (b) Define two matrices A and B to be *congruent* if there exists a nonsingular matrix P such that $P^T A P = B$. Show that this is an equivalence relation.

3. Let A and B be similar matrices. Prove or disprove:
   (a) A^T is similar to B^T.
   (b) A^n is similar to B^n if n is a positive integer.
   (c) If A is invertible, then B is invertible.

4. Let A be an n by n matrix with n distinct eigenvalues $k_1, \ldots, k_n$.
   (a) Explain why A is similar to any other matrix with the same eigenvalues.
   (b) Explain why $\det A = k_1 k_2 \ldots k_n$ by comparing A with an appropriate matrix $P^{-1}AP$.

5. (a) Give an example of two matrices which have the same eigenvalues but are not similar.
   (b) Prove or disprove: If A_1 is similar to B_1, and A_2 is similar to B_2, then $A_1 A_2$ is similar to $B_1 B_2$.

6. Let A and B be similar. Show:
   (a) A and B have the same rank. Recall that rank $MN \leq$ min {rank M, rank N}. That is, the rank of the product of two matrices cannot exceed the rank of either factor.
   (b) Trace (A) = Trace (B). (Recall that Trace A is equal to the sum of its eigenvalues, from Lesson 17, Exercise 9.)

7. Explain why the product of (upper) triangular matrices is triangular.

8. Given $A = \begin{bmatrix} 0 & -3 & 4 \\ -3 & -10 & 14 \\ -1 & -6 & 8 \end{bmatrix}$, display a matrix P such that $P^{-1}AP$ is diagonal.

9. Use Exercise 8 to solve the system of differential equations

$$\frac{dy_1}{dx} = -3y_2 + 4y_3$$

$$\frac{dy_2}{dx} = -3y_1 - 10y_2 + 14y_3$$

$$\frac{dy_3}{dx} = -y_1 - 6y_2 + 8y_3$$

10. If P is a constant n by n matrix and $Y = (y_1, y_2, \ldots, y_n)$, where each y_i is a differentiable function of x, show that

$$\frac{d}{dx}(PY^T) = P\frac{dY^T}{dx}$$

Similarity and Canonical Forms

11. Let $A = \begin{bmatrix} 1 & 4 \\ -1 & 5 \end{bmatrix}$. Find P so that $P^{-1}AP$ is triangular. Explain why A is not similar to a diagonal matrix.

12. Partition the following set of matrices into similar classes:
I, $A = \begin{bmatrix} 1 & 1 \\ 0 & 1 \end{bmatrix}$, $B = \begin{bmatrix} 1 & 0 \\ 0 & -1 \end{bmatrix}$, $C = \begin{bmatrix} 0 & 1 \\ 1 & 0 \end{bmatrix}$, $D = \begin{bmatrix} 1 & 2 \\ 0 & 1 \end{bmatrix}$,
$E = \begin{bmatrix} 3 & -4 \\ 2 & -3 \end{bmatrix}$.

13. By an appropriate partitioning of matrices, find
$$\begin{bmatrix} 1 & 0 & 0 & 0 & 0 \\ 0 & 1 & 0 & 0 & 0 \\ 0 & 0 & 1 & 0 & 0 \\ 0 & 0 & 0 & 2 & 3 \\ 0 & 0 & 0 & 1 & 2 \end{bmatrix} \begin{bmatrix} a_{11} & a_{12} & a_{13} & a_{14} & a_{15} \\ a_{21} & a_{22} & a_{23} & a_{24} & a_{25} \\ a_{31} & a_{32} & a_{33} & a_{34} & a_{35} \\ 2 & -3 & 0 & 1 & 1 \\ -1 & 2 & 0 & 1 & 1 \end{bmatrix}$$

14. Find the inverse of
$$\begin{bmatrix} 2 & 0 & 0 & 0 & 0 & 0 & 0 \\ 0 & 1 & 0 & 0 & 0 & 0 & 0 \\ 0 & 0 & 3 & 2 & 0 & 0 & 0 \\ 0 & 0 & 4 & 3 & 0 & 0 & 0 \\ 0 & 0 & 0 & 0 & 0 & 0 & 1 \\ 0 & 0 & 0 & 0 & 0 & 1 & 0 \\ 0 & 0 & 0 & 0 & 1 & 0 & 0 \end{bmatrix}$$

15. Let T be an orthogonal linear transformation of R^n into R^n, that is, let T be a linear transformation whose matrix relative to an orthonormal basis is the orthogonal matrix P. If k is a real eigenvalue of T, show from the definition of eigenvector that $|k| = 1$. From this conclude that each real eigenvalue of an orthogonal matrix has absolute value 1, and that an orthogonal matrix of odd order has either 1 or -1 as an eigenvalue. Remark: It is also true that the imaginary eigenvalues of an orthogonal matrix have modulus 1.

16. Use matrix algebra and the theory of determinants to show:
    (a) If P is proper orthogonal of odd order, then 1 is an eigenvalue of P. Hint: Show $|P - I| = 0$.
    (b) If P is improper orthogonal, then -1 is an eigenvalue of P.

17. Find the eigenvalues of each of the following orthogonal matrices, and check that each eigenvalue has absolute value 1.
    (a) $\frac{1}{5}\begin{bmatrix} 3 & 4 \\ 4 & -3 \end{bmatrix}$;
    (b) $\frac{1}{5}\begin{bmatrix} 3 & 4 \\ -4 & 3 \end{bmatrix}$;

Lessons in Linear Algebra

(c) $\dfrac{1}{3}\begin{bmatrix} 2 & 2 & -1 \\ 1 & -2 & -2 \\ 2 & -1 & 2 \end{bmatrix}$; (d) $\dfrac{1}{3}\begin{bmatrix} 2 & 2 & 1 \\ 1 & -2 & 2 \\ 2 & -1 & -2 \end{bmatrix}$.

18. Let $T: R^2 \to R^2$, $T(x, y) = (2x - y, 2x + y)$, and let $\alpha = \{(1, 1), (1, -1)\}$, $\mathcal{B} = \{(1, 2), (1, 0)\}$ be two bases.
    (a) Find P, the matrix of transition from α to $\mathcal{B}$.
    (b) Find A and B, the matrices representing T relative to the bases α and $\mathcal{B}$, respectively.
    (c) Let $\alpha = (4, 3)$. Find the coordinate X of α relative to α, and find the coordinate X' of α relative to $\mathcal{B}$.
    (d) Find the coordinates Y and Y' of $T(\alpha)$ relative to α and to $\mathcal{B}$.

19. (a) Let $A = \begin{bmatrix} 29 & 12 \\ 12 & 36 \end{bmatrix}$. Find a proper orthogonal matrix P such that $P^{-1}AP$ is diagonal.
    (b) Let $X = [x \ \ y]$ and consider the conic section with equation $XAX^T = 180$. Use P as a matrix of transition from one basis to another and find the new equation of the conic section. Sketch.

Remarks on Exercises

1. The eigenvalues of A are $\pm 2i$, so A cannot be similar to a real diagonal matrix. Let $P = \begin{bmatrix} 5 & 5 \\ 1 - 2i & 1 + 2i \end{bmatrix}$. Then $P^{-1}AP = \begin{bmatrix} 2i & 0 \\ 0 & -2i \end{bmatrix}$. P is not unique.

3. All are true. For (b) use $(P^{-1}AP)^n = P^{-1}A^nP$.

4. (a) Since A has distinct eigenvalues, A is similar to
$$B = P^{-1}AP = \begin{bmatrix} k_1 & & & \bigcirc \\ & k_2 & & \\ & & \ddots & \\ \bigcirc & & & k_n \end{bmatrix}$$
Also $|A| = |B| = k_1 \ldots k_n$.

5. (a) $\begin{bmatrix} 1 & 0 \\ 0 & 1 \end{bmatrix}$ and $\begin{bmatrix} 1 & 1 \\ 0 & 1 \end{bmatrix}$. I is similar only to I.
   (b) False. For a counterexample, let $A_1 = \begin{bmatrix} 1 & 0 \\ 0 & 2 \end{bmatrix}$, $B_1 = \begin{bmatrix} 0 & 2 \\ -1 & 3 \end{bmatrix}$,

Similarity and Canonical Forms

$A_2 = B_2 = \begin{bmatrix} 0 & 1 \\ 1 & 0 \end{bmatrix}$. Then A_1 is similar to B_1, A_2 is similar to B_2, but $A_1 A_2$ is not similar to $B_1 B_2$.

7. Let $A = [a_{ij}]$, $B = [b_{ij}]$ be upper triangular. This means that $a_{ij} = 0$ and $b_{ij} = 0$ whenever $i > j$. Now we will show that the ij-element of AB, namely $\sum a_{ik} b_{kj}$, is 0 whenever $i > j$. It turns out that each individual summand $a_{ik} b_{kj}$ is 0 whenever $i > j$, because if $k > j$, then $b_{kj} = 0$, and if $j \geq k$, then $i > k$, and $a_{ik} = 0$.

8. The characteristic polynomial is $-x^3 - 2x^2 + x + 2$, and the eigenvalues are 1, -1, and -2. We may take $P = \begin{bmatrix} 1 & 3 & 2 \\ 1 & 13 & 8 \\ 1 & 9 & 5 \end{bmatrix}$ and get $P^{-1}AP = \begin{bmatrix} 1 & 0 & 0 \\ 0 & -1 & 0 \\ 0 & 0 & -2 \end{bmatrix}$.

9. $y_1 = c_1 e^x + 3c_2 e^{-x} + 2c_3 e^{-2x}$, $y_2 = c_1 e^x + 13 c_2 e^{-x} + 8 c_3 e^{-2x}$,
$y_3 = c_1 e^x + 9 c_2 e^{-x} + 5 c_3 e^{-2x}$, or $Y^T = P \begin{bmatrix} c_1 e^x \\ c_2 e^{-x} \\ c_3 e^{-2x} \end{bmatrix}$.

10. $\dfrac{d}{dx}(PY^T) = \dfrac{d}{dx} \begin{bmatrix} \sum p_{1k} y_k \\ \sum p_{2k} y_k \\ \vdots \\ \sum p_{nk} y_k \end{bmatrix} = \begin{bmatrix} \sum p_{1k} y'_k \\ \vdots \\ \sum p_{nk} y'_k \end{bmatrix} = P \dfrac{dY^T}{dx}$.

11. The eigenvalues are equal, $k_1 = k_2 = 3$, and the eigenspace is of dimension 1. Since there is no set of two independent eigenvectors, A is not similar to a diagonal matrix. Take the eigenvector $(2, 1)$ and construct a nonsingular matrix P with $(2, 1)$ in the first column. Let $P = \begin{bmatrix} 2 & 0 \\ 1 & 1 \end{bmatrix}$. Then $P^{-1}AP = \begin{bmatrix} 3 & 2 \\ 0 & 3 \end{bmatrix}$. The answer is not unique. Take $Q = \begin{bmatrix} 2 & 1 \\ 1 & 1 \end{bmatrix}$. Then $Q^{-1}AQ = \begin{bmatrix} 3 & 1 \\ 0 & 3 \end{bmatrix}$.

Lessons in Linear Algebra

12. I is similar only to itself. A and D are similar, because with $P = \begin{bmatrix} 1 & 0 \\ 0 & 2 \end{bmatrix}$ we have $P^{-1}AP = D$. Neither A nor D is similar to a diagonal matrix because neither has two independent eigenvectors. B, C, E are similar, because each has two distinct eigenvalues $1, -1$ and is therefore similar to $\begin{bmatrix} 1 & 0 \\ 0 & -1 \end{bmatrix}$.

15. Let $T(\alpha) = k\alpha$. Since T is orthogonal, it preserves inner products, and we have $(T(\alpha), T(\alpha)) = k^2(\alpha, \alpha) = (\alpha, \alpha)$, so $k^2 = 1$. An orthogonal matrix of odd order has a characteristic equation of odd order, and therefore must have a real eigenvalue, necessarily ± 1.

16. (a) Since P is proper orthogonal, $|P| = |P^T| = 1$. Therefore, $|P - I| = |P^T| |P - I| = |I - P| = (-1)^{\text{odd}} |P - I| = -|P - I|$, and $|P - I| = 0$.

17. (a) ± 1; (b) $\frac{1}{5}(3 \pm 4i)$;

    (c) The characteristic equation is $3x^3 - 2x^2 - 2x + 3 = 0$. The eigenvalues are $-1, \frac{1}{6}(5 \pm i\sqrt{11})$.

    (d) Characteristic equation: $3x^3 + 2x^2 - 2x - 3 = 0$. Eigenvalues: $1, \frac{1}{6}(-5 \pm i\sqrt{11})$.

18. (a) $P = \frac{1}{2}\begin{bmatrix} 3 & 1 \\ -1 & 1 \end{bmatrix}$.

    (b) $A = \begin{bmatrix} 2 & 2 \\ -1 & 1 \end{bmatrix}$; $B = \begin{bmatrix} 2 & 1 \\ -2 & 1 \end{bmatrix} = P^{-1}AP$.

    (c) $X = \begin{bmatrix} 7 & 1 \\ 2 & 2 \end{bmatrix}$, $X' = \begin{bmatrix} 3 & 5 \\ 2 & 2 \end{bmatrix}$.

    (d) $Y = [8 \ \ -3]$, $Y' = \begin{bmatrix} \frac{11}{2} & -\frac{1}{2} \end{bmatrix}$.
    $Y^T = AX^T$, $(Y')^T = B(X')^T$.

19. (a) The eigenvalues of A are 20 and 45. We use unit eigenvectors parallel to the eigenvectors $(4, -3)$ and $(3, 4)$ to construct $P = \frac{1}{5}\begin{bmatrix} 4 & 3 \\ -3 & 4 \end{bmatrix}$, a proper orthogonal matrix. Then $P^{-1}AP = \begin{bmatrix} 20 & 0 \\ 0 & 45 \end{bmatrix}$.

    (b) The matrix equation $XAX^T = 180$ may be written $29x^2 + 24xy + 36y^2 = 0$, representing a conic section relative to a

Similarity and Canonical Forms

rectangular coordinate system. Let P represent a change of basis (coordinates) from $\alpha = \{(1, 0), (0, 1)\}$ to $\mathcal{B} = \left\{\left(\frac{4}{5}, -\frac{3}{5}\right), \left(\frac{3}{5}, \frac{4}{5}\right)\right\}$. This is a rotation of axes. The new coordinates $X' = (x', y')$ of a point are given by $X^T = P(X')^T$, so the new equation of the conic is $X'P^T AP(X')^T = 180$, $X'P^{-1}AP(X')^T = 180$, or

$$[x' \ y'] \begin{bmatrix} 20 & 0 \\ 0 & 45 \end{bmatrix} \begin{bmatrix} x' \\ y' \end{bmatrix} = 180$$

This gives

$$\frac{x'^2}{9} + \frac{y'^2}{4} = 1,$$

an ellipse (see Figure 18-2).

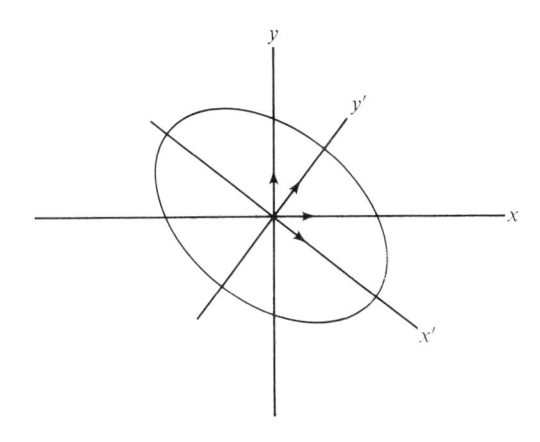

FIGURE 18-2

233

19

APPLICATIONS TO GEOMETRY

Applications of linear algebra to various problems in plane and solid analytic geometry are described in this lesson. We represent planes and lines with equations, and we examine the relationships among points, lines, and planes.

We use projections to solve several problems. We find the distance between a plane and a point, the distance between a line and a point, the distance between two parallel planes, and the distance between two skew-lines.

The idea of projection is generalized and it is proved that the linear transformation $T: V \to V$ is a projection if and only if T is *idempotent*, that is, if and only if $T^2 = T$.

The lesson is concluded with an introduction to the *vector product* (or *cross-product*) $\alpha \times \beta$ of two vectors. The cross-product is interpreted in terms of areas and volumes.

19-1 Theorem

The equation of the plane through the point $X_0 = (x_0, y_0, z_0)$ with normal vector $N = (a, b, c) \neq 0$ is $NX^T = NX_0^T$ or $ax + by + cz = ax_0 + by_0 + cz_0$.

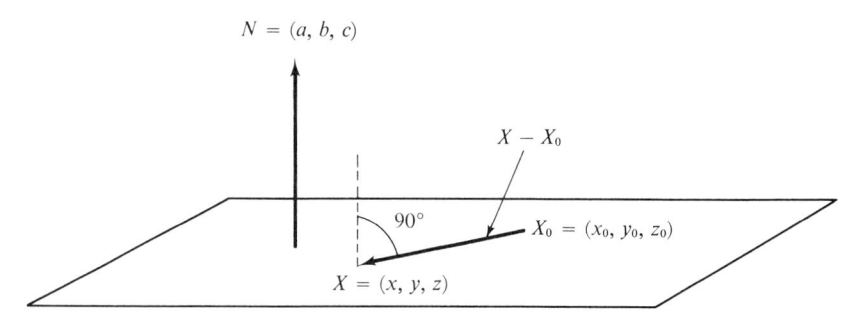

FIGURE 19-1

Applications to Geometry

Proof: See Figure 19-1. The variable point $X = (x, y, z)$ of three-space is on the plane if and only if the vector $X - X_0$ is orthogonal to N, or if and only if

$$N(X - X_0)^T = 0 \quad \blacksquare$$

Sometimes a, b, c are called "direction numbers" of the normal.

Example 1. We find the equation of the plane through $(3, 1, -2)$ perpendicular to the vector $(1, -2, 5)$.

$$(1, -2, 5) \cdot (x, y, z) = (1, -2, 5) \cdot (3, 1, -2)$$

or

$$x - 2y + 5z + 9 = 0$$

Example 2. We sketch the planes $2x + 3y + 4z - 12 = 0$ and $x - y + 1 = 0$. (See Figure 19-2.) The first plane cuts through the coordinate

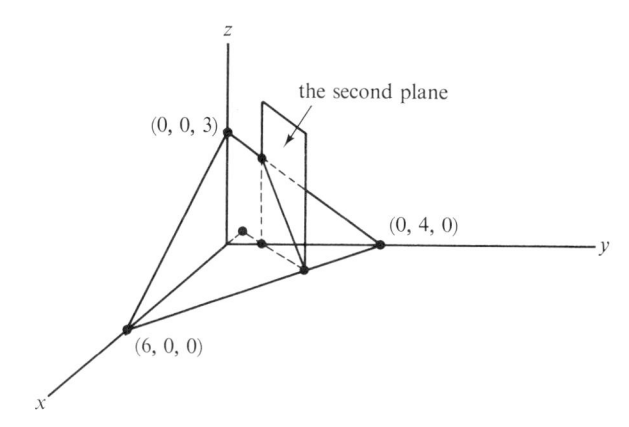

FIGURE 19-2

axes at the points $(6, 0, 0)$, $(0, 4, 0)$, $(0, 0, 3)$. The numbers 6, 4, and 3 are called the "intercepts" of the plane. The second plane, being normal to the vector $(1, -1, 0)$, is parallel to the z-axis and contains the points $(-1, 0, 0)$ and $(0, 1, 0)$. By solving the two equations simultaneously we find that they meet at points $(9/5, 14/5, 0)$ and $(0, 1, 9/4)$.

The points satisfying simultaneously the equations of two planes are those points on the line of intersection of the two planes. The equations of a line can be written in several equivalent forms.

Lessons in Linear Algebra

19-2 Theorem

The point $X = (x, y, z)$ is on the line through $X_0 = (x_0, y_0, z_0)$ parallel to the vector $A = (a, b, c)$ if and only if $X = X_0 + tA$ for some real number t.

Proof: See Figure 19-3. The vector $X - X_0$ is parallel to A if and only if $X - X_0 = $ a multiple of A, for example tA.

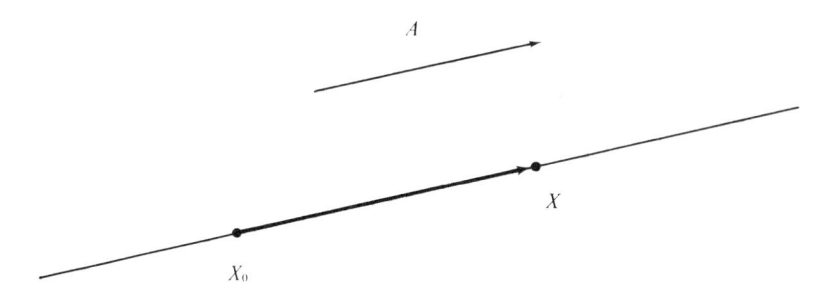

FIGURE 19-3

The equation $X = X_0 + tA$ is the *vector* equation of the line. If we write the vector equation out component-wise we get

$$x = x_0 + at$$
$$y = y_0 + bt$$
$$z = z_0 + ct$$

the *parametric equations* of the line. The idea is that $(x - x_0, y - y_0, z - z_0)$ is proportional to the direction numbers (a, b, c). This is sometimes indicated by

$$\frac{x - x_0}{a} = \frac{y - y_0}{b} = \frac{z - z_0}{c}$$

the *symmetric* form of the equations of the line, but of course if a or b or c is zero a special interpretation must be made. ■

Example 3. We find the equations of the line through $(1, 3, -4)$ parallel to the vector $(2, 1, 5)$.

The vector equation of the line is

$$(x, y, z) = (1, 3, -4) + t(2, 1, 5)$$

The parametric equations are

$$x = 1 + 2t$$
$$y = 3 + t$$
$$z = -4 + 5t$$

The symmetric form is $(x - 1)/2 = (y - 3)/1 = (z + 4)/5$.

These equations are by no means unique. We could use any point on the line in place of the point $(1, 3, -4)$.

Example 4. Let us find parametric equations of the line of intersection of the planes $x + y - 2z + 1 = 0$, $2x - y + z - 4 = 0$.

We need two things: (1) a point on the line and (2) a vector A parallel to the line. The first is easy to obtain. Take $X = (0, -7, -3)$. The vector $A = (a, b, c)$ must be perpendicular to the normal of each plane. That is $(a, b, c) \cdot (1, 1, -2) = 0$, and $(a, b, c) \cdot (2, -1, 1) = 0$. Therefore, we choose $A = (1, 5, 3)$ and the parametric equations of the required line are

$$x = t$$
$$y = -7 + 5t$$
$$z = -3 + 3t$$

Example 5. Find where the line through the points $X_1 = (1, 3, 4)$ and $X_2 = (2, -5, 2)$ meets the planes $x - 2y + z - 5 = 0$ and $4x + y - 2z + 3 = 0$.

Solution: The direction numbers of the line are given by the vector $X_2 - X_1 = (1, -8, -2)$. Hence, equations of the line are $x = 1 + t$, $y = 3 - 8t$, $z = 4 - 2t$.

The line meets the plane $x - 2y + z - 5 = 0$ when $(1 + t) - 2(3 - 8t) + (4 - 2t) - 5 = 0$ or when $t = 2/5$. This produces the point $(7/5, -1/5, 16/5)$.

The line meets the plane $4x + y - 2z + 3 = 0$ when $4(1 + t) + (3 - 8t) - 2(4 - 2t) + 3 = 0$ or when $2 = 0$. We conclude that the line does not meet the second plane.

In Lessons 7 and 8 we treated orthogonality in Euclidean spaces. Let us apply these ideas to the geometry of three-space. To project a point P upon a line (plane) we draw a perpendicular from P to the line (plane). The foot of the perpendicular is the projection of P upon the line (plane). Projections of sets of points (for example, lines, line-segments, curves,

Lessons in Linear Algebra

disks, spheres, etc.) are defined point-wise. We say the projection of the set S upon a plane π is the set of points in π which are projections of points in S.

Example 6. Let f be the projection which sends every point X in three-space into its projection on the plane π (see Figure 19-4). The projection

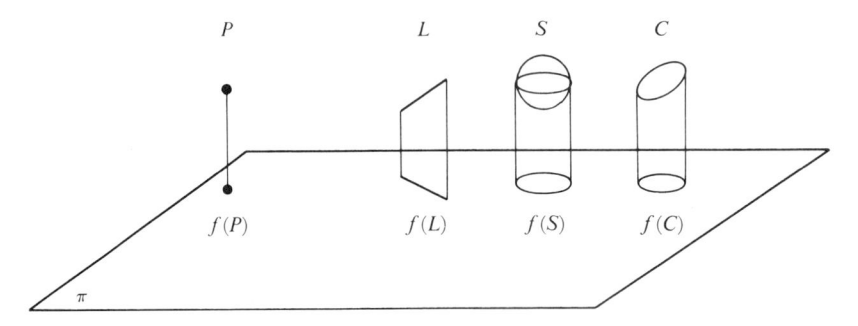

FIGURE 19-4

of a line-segment is a line-segment, the projection of a sphere is a circle, etc.

Example 7. Now let g be the projection which sends each point X in three-space into its projection on the line L (see Figure 19-5). The pro-

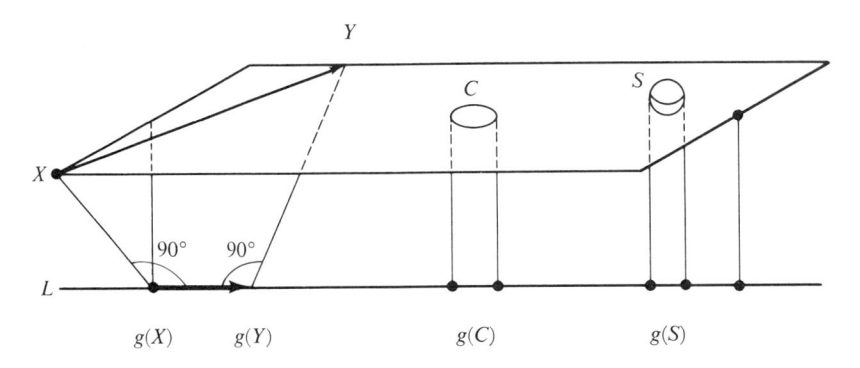

FIGURE 19-5

jection of the vector from X to Y is the vector from $g(X)$ to $g(Y)$. The projection of a sphere S is a line segment $g(S)$, etc.

Applications to Geometry

Knowing how to find the length of the projection of a line segment XY upon a plane or line permits us to handle a number of geometric problems.

Two geometric vectors may be equal but have different initial points. Even so their projections on a line (plane) are equal. We recall some earlier results.

Let A be any vector in three-space. The projection of A upon the vector B is $[(A \cdot B)/(B \cdot B)]B$. The projection of A upon the plane π is $A - [(A \cdot N)/(N \cdot N)]N$, where N is a vector normal to π.

19-3
Theorem

We will not repeat the proof here, but you can reconstruct it by looking at Figure 19-6.

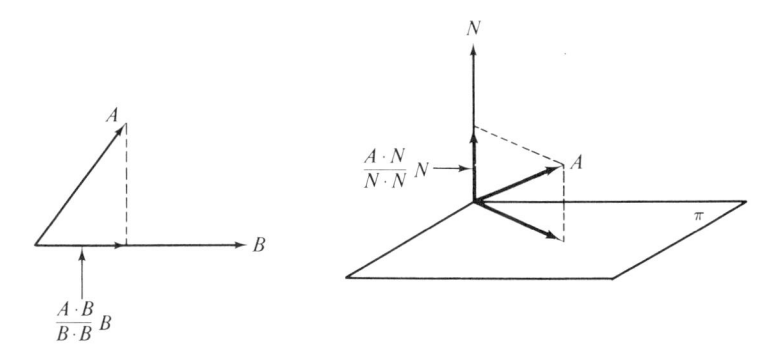

FIGURE 19-6

Common geometric problems are those of finding distances between a plane and a point, between a line and a point, between skew-lines, etc. These can be handled efficiently by using the idea of projections. It is unnecessary and perhaps sinful to memorize deliberately intricate methods for each problem.

We illustrate a plan for handling each of several problems.

Example 8. *To find the distance from a plane π to a point X:* See Figure 19-7. Plan: Let Y be any point on π and find the length of the projection of $\overrightarrow{XY}$ upon N, a normal to the plane.

To illustrate we find the distance from the plane $3x - 4y + z - 5 = 0$ to the point $X = (1, 1, 3)$. A vector normal to the plane is $N = (3, -4, 1)$. We choose a point $Y = (1, 0, 2)$ on the plane. The projection of $\overrightarrow{XY} =$

Lessons in Linear Algebra

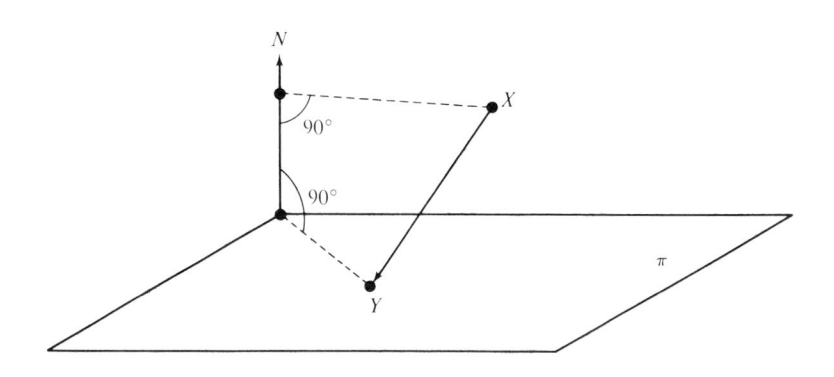

FIGURE 19-7

$(0, -1, -1)$ upon N is $(3/26)(3, -4, 1)$. The distance from the plane to the point is $N = 3/\sqrt{26}$.

Note: In the past you may have used the following fact: The distance from the plane $ax + by + cz + d = 0$ to the point (x_0, y_0, z_0) is $|ax_0 + by_0 + cz_0 + d|/\sqrt{a^2 + b^2 + c^2}$. It is left as an exercise for you to prove the formula valid.

Example 9. To find the distance from a line L to a point X: See Figure 19-8. Plan: Choose a point Y on L and find the length of the projection of $\overrightarrow{XY}$ upon a plane perpendicular to L.

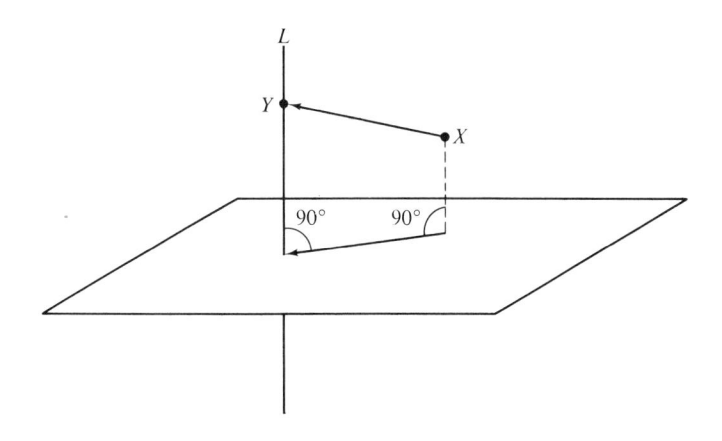

FIGURE 19-8

Applications to Geometry

In two space we find the distance from the line $3x + 4y - 12 = 0$ to the point $X = (1, 2)$ by choosing the point $Y = (0, 3)$ on the line and finding the projection of $\overrightarrow{XY} = (-1, 1)$ upon $N = (3, 4)$, a vector perpendicular to the line. The required distance is the length of $(1/25)(3, 4)$ or $1/5$.

The problem in three-space is handled the same way. Let us find the distance d from the line through the points $A(1, 1, 2)$ and $B(3, 0, 5)$ to the point $X(10, -3, 4)$.

The projection of $\overrightarrow{AX}$ upon a plane normal to AB is

$$\overrightarrow{AX} - \frac{\overrightarrow{AX} \cdot \overrightarrow{AB}}{\overrightarrow{AB} \cdot \overrightarrow{AB}} \overrightarrow{AB} = (9, -4, 2) - \frac{(9, -4, 2) \cdot (2, -1, 3)}{(2, -1, 3) \cdot (2, -1, 3)} (2, -1, 3)$$
$$= (9, -4, 2) - 2(2, -1, 3) = (5, -2, -4)$$

and the required distance $d = 3\sqrt{5}$.

Example 10. *To find the distance between two parallel planes:* See Figure 19-9. Plan: Choose two points X and Y, one from the first plane, the other

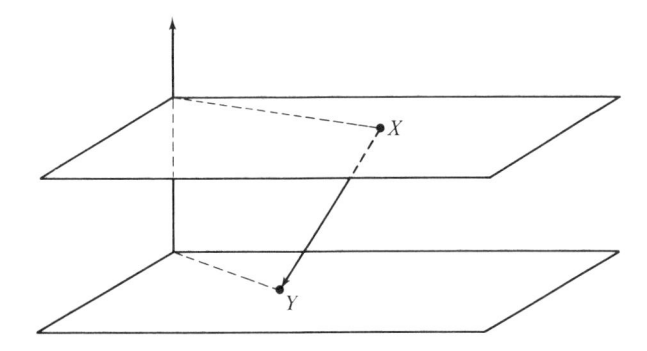

FIGURE 19-9

from the second. Then project $\overrightarrow{XY}$ upon a vector normal to the two planes. (This is the same as finding the distance from one plane to any point on the other.)

Example 11. *To find the distance between two skew-lines:* See Figure 19-10. Plan: Find the distance between two parallel planes containing the given skew-lines.

To illustrate, let us find the distance between lines AB and CD, given $A(1, 1, 2)$, $B(-3, 0, 5)$, $C(1, 2, 2)$, $D(-1, 1, 0)$. First, we find a vector

241

Lessons in Linear Algebra

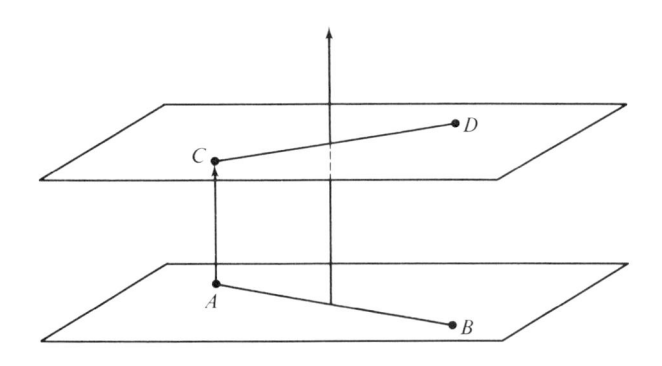

FIGURE 19-10

normal to $\vec{AB} = (-4, -1, 3)$ and $\vec{CD} = (-2, -1, -2)$. The vector is found to be $N = (5, -14, 2)$. The projection of $\vec{AC} = (0, 1, 0)$ upon N is $[(\vec{AC} \cdot N)/(N \cdot N)]N = (-14/225)(5, -14, 2)$, so the required distance is $14/15$.

In all of the above examples we have used projections, very special linear transformations which have unusual properties. Consider the projection $T : R^3 \to R^3$ which sends each point into its projection on the xy-plane. (The rule is $T(x, y, z) = (x, y, 0)$.) Observe that $T^2 = T$; that is, T is "idempotent." Also, T is the identity mapping on its image space, and the kernel of T is the orthogonal complement of the image space of T.

A lovely result is that idempotency is the algebraic earmark of a linear transformation which is a projection (not necessarily orthogonal). We generalize to spaces of higher dimension.

19-4 Definition

Let the linear space V be the direct sum of two subspaces S_1 and S_2; that is, let $V = S_1 + S_2$, where $S_1 \cap S_2 = 0$. If $\alpha = \alpha_1 + \alpha_2$, with $\alpha_1 \varepsilon S_1$, $\alpha_2 \varepsilon S_2$, then α_1, α_2 are called the *components* of α in S_1, S_2, respectively.

The linear transformation T for which $T(\alpha) = \alpha_1$ is called the *projection of V upon S_1 along S_2*.

It is clear that the components α_1, α_2 of any vector are unique because if there were two such representations $\alpha_1 + \alpha_2 = \beta_1 + \beta_2$, we would have $\alpha_1 - \beta_1 = \beta_2 - \alpha_2$, a vector of $S_1 \cap S_2$, and hence $\alpha_1 - \beta_1 = \beta_2 - \alpha_2 = 0$.

Applications to Geometry

19-5
Theorem

Let T be a linear transformation of V into V. Then T is a projection if and only if $T^2 = T$. In this case T is the projection of V upon the image space of T, along the kernel of T.

Proof: First suppose T is the projection of V upon S_1 along S_2. Then for $\alpha = \alpha_1 + \alpha_2$, $\alpha_1 \varepsilon S_1$, $\alpha_2 \varepsilon S_2$, we have $T^2(\alpha) = T(T(\alpha)) = T(\alpha_1) = \alpha_1 = T(\alpha)$, and hence $T^2 = T$.

Conversely, suppose $T^2 = T$, let $S_1 = T(V)$, the image space of T, and let $S_2 = $ kernel (T). We need to show V is the direct sum of S_1 and S_2. The general vector $\alpha = T(\alpha) + (\alpha - T(\alpha))$, where $T(\alpha) \varepsilon S_1$ and $\alpha - T(\alpha) \varepsilon S_2$ since $T(\alpha - T(\alpha)) = T(\alpha) - T^2(\alpha) = T(\alpha) - T(\alpha) = 0$. Also, $S_1 \cap S_2 = \{0\}$, because $\beta \varepsilon S_1 \cap S_2$ would imply $\beta = T(\alpha)$ for some α, and $T(\beta) = 0$. Then we would have $0 = T(\beta) = T^2(\alpha) = T(\alpha) = \beta$. Therefore, T is the projection of V upon $T(V)$ along the kernel of V. ∎

Example 12. Let $T : R^3 \to R^3$ by the rule $T(x, y, z) = (x, x, z)$. Clearly $T_2 = T$, so T is a projection. The image space is $S_1 = \{x(1, 1, 0) + z(0, 0, 1) \mid x, z \varepsilon R\}$, the plane $x = y$. The kernel is the y-axis. Therefore, T is the projection of V upon the plane $x = y$, along the y-axis. (See Figure 19-11.)

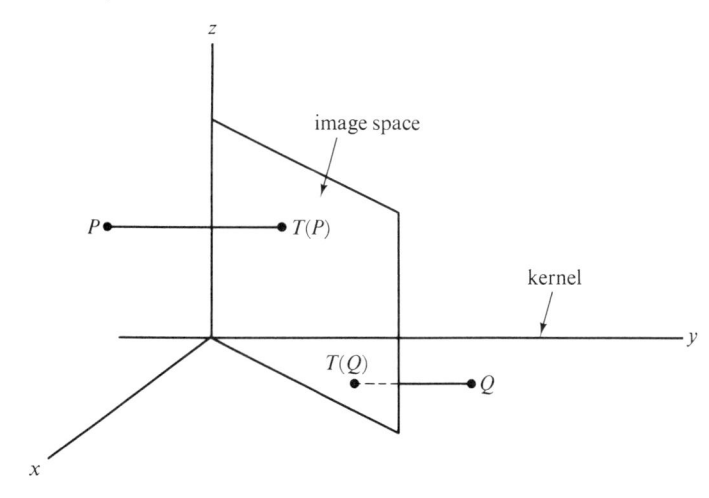

FIGURE 19-11

The "vector product" of two vectors in three-space is used frequently by engineers and scientists. We conclude this lesson by introducing the vector product and interpreting it in terms of areas and volumes.

Lessons in Linear Algebra

19-6 Definition

Let $\alpha = (a_1, a_2, a_3)$, $\beta = (b_1, b_2, b_3)$ be two vectors in R^3. The *vector product* (or *cross-product*) of α and β, denoted $\alpha \times \beta$ is defined by

$$\alpha \times \beta = \left(\begin{vmatrix} a_2 & a_3 \\ b_2 & b_3 \end{vmatrix}, \begin{vmatrix} a_3 & a_1 \\ b_3 & b_1 \end{vmatrix}, \begin{vmatrix} a_1 & a_2 \\ b_1 & b_2 \end{vmatrix} \right)$$

(Note: This can be recalled by means of the "determinant"

$$\begin{vmatrix} i & j & k \\ a_1 & a_2 & a_3 \\ b_1 & b_2 & b_3 \end{vmatrix}$$

where i, j, k are the usual unit vectors, $i = (1, 0, 0)$, $j = (0, 1, 0)$, $k = (0, 0, 1)$.)

It is easy to observe that the vector $\alpha \times \beta$ is perpendicular to both α and β, because $\alpha \times \beta \cdot \beta = 0$ and $\alpha \times \beta \cdot \alpha = 0$. Also, $\alpha \times \beta$ can be interpreted in terms of area.

19-7 Theorem

Let $\alpha = (a_1, a_2, a_3)$, $\beta = (b_1, b_2, b_3)$. Then
(a) $\alpha \times \beta = -\beta \times \alpha$.
(b) $\alpha \times \beta$ is perpendicular to both α and β.
(c) $\|\alpha \times \beta\|$ is the area of the parallelogram formed by α and β.

Proof: Parts (a) and (b) are left as exercises.

(c) The area of the parallelogram is $\|\alpha\| \|\beta\| \sin \theta$, where θ is the angle formed by α and β. Hence,

$$(\text{Area})^2 = \|\alpha\|^2 \|\beta\|^2 [1 - \cos^2 \theta]$$
$$= \|\alpha\|^2 \|\beta\|^2 \left[1 - \left(\frac{\alpha \cdot \beta}{\|\alpha\| \|\beta\|} \right)^2 \right]$$
$$= \|\alpha\|^2 \|\beta\|^2 - (\alpha \cdot \beta)^2$$
$$= (a_1^2 + a_2^2 + a_3^2)(b_1^2 + b_2^2 + b_3^2) - (a_1 b_1 + a_2 b_2 + a_3 b_3)^2$$
$$= a_2^2 b_3^2 + a_3^2 b_2^2 + a_3^2 b_1^2 + a_1^2 b_3^2 + a_1^2 b_2^2 + a_2^2 b_1^2 - 2a_2 b_2 a_3 b_3$$
$$\quad - 2a_3 b_3 a_1 b_1 - 2a_1 b_1 a_2 b_2$$
$$= (a_2 b_3 - a_3 b_2)^2 + (a_3 b_1 - a_1 b_3)^2 + (a_1 b_2 - a_2 b_1)^2$$
$$= \|\alpha \times \beta\|^2 \quad \blacksquare$$

19-8 Theorem

The volume of the parallelepiped formed by the vectors α, β, γ is the absolute value of $\alpha \cdot \beta \times \gamma$, the "triple scalar product" of α, β, γ.

Proof: The area of the base formed by β, γ is $\|\beta \times \gamma\|$, and the corresponding altitude is the length of projection of α upon $\beta \times \gamma$, namely

Applications to Geometry

$|\alpha \cdot \beta \times \gamma|/\|\beta \times \gamma\|$. (See Figure 19-12.) Multiplying the base area times the altitude completes the proof. ∎

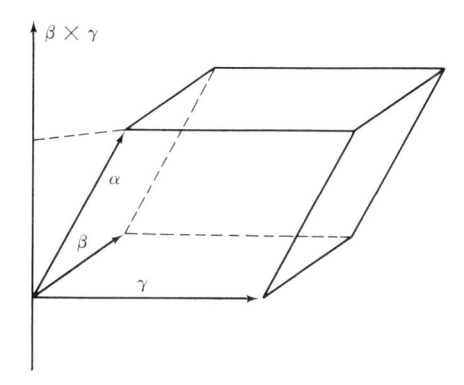

FIGURE 19-12

Let $\alpha = (a_1, a_2, a_3)$, $\beta = (b_1, b_2, b_3)$, $\gamma = (c_1, c_2, c_3)$. Then

$$\alpha \cdot \beta \times \gamma = \alpha \times \beta \cdot \gamma = \begin{vmatrix} a_1 & a_2 & a_3 \\ b_1 & b_2 & b_3 \\ c_1 & c_2 & c_3 \end{vmatrix}$$

19-9
Theorem

Proof:

$$\alpha \cdot \beta \times \gamma = (a_1, a_2, a_3) \cdot \left(\begin{vmatrix} b_2 & b_3 \\ c_2 & c_3 \end{vmatrix}, \begin{vmatrix} b_3 & b_1 \\ c_3 & c_1 \end{vmatrix}, \begin{vmatrix} b_1 & b_2 \\ c_1 & c_2 \end{vmatrix} \right)$$

$$= \begin{vmatrix} a_1 & a_2 & a_3 \\ b_1 & b_2 & b_3 \\ c_1 & c_2 & c_3 \end{vmatrix}$$

This is apparent when the determinant is evaluated by using the cofactors of elements from the first row.

Finally,

$$\alpha \times \beta \cdot \gamma = \gamma \cdot \alpha \times \beta \qquad \text{(dot product is commutative)}$$

$$= \begin{vmatrix} c_1 & c_2 & c_3 \\ a_1 & a_2 & a_3 \\ b_1 & b_2 & b_3 \end{vmatrix} \qquad \text{(from the rule just proved)}$$

$$= \begin{vmatrix} a_1 & a_2 & a_3 \\ b_1 & b_2 & b_3 \\ c_1 & c_2 & c_3 \end{vmatrix} \qquad \text{(properties of determinants)}$$

$$= \alpha \cdot \beta \times \gamma \qquad \blacksquare$$

Lessons in Linear Algebra

Example 13. Let $\alpha = (1, 0, 2)$, $\beta = (-1, -1, 1)$, $\gamma = (1, 2, 4)$. Then

$$\beta \times \gamma = \left(\begin{vmatrix} -1 & 1 \\ 2 & 4 \end{vmatrix}, \begin{vmatrix} 1 & -1 \\ 4 & 1 \end{vmatrix}, \begin{vmatrix} -1 & -1 \\ 1 & 2 \end{vmatrix} \right) = (-6, 5, -1)$$

and we note that $\beta \times \gamma$ is orthogonal to both β and γ. $\|\beta \times \gamma\| = \sqrt{62}$, the area of the parallelogram formed by β and γ.

Also, $\alpha \cdot \beta \times \gamma = (1, 0, 2) \cdot (-6, 5, -1) = -8$, or equivalently

$$\alpha \cdot \beta \times \gamma = \begin{vmatrix} 1 & 0 & 2 \\ -1 & -1 & 1 \\ 1 & 2 & 4 \end{vmatrix} = -8.$$ Therefore, the volume of the parallelepiped formed by α, β, γ is 8.

Summary of Topics in Lesson 19

The equation of the plane through (x_0, y_0, z_0) normal to (a, b, c) is $ax + by + cz = ax_0 + by_0 + cz_0$.

The equations of the line through (x_0, y_0, z_0) with direction numbers (a, b, c) are

$$\left. \begin{array}{l} x = x_0 + at \\ y = y_0 + bt \\ z = z_0 + ct \end{array} \right\} \quad \text{(parametric form)}$$

$$\frac{x - x_0}{a} = \frac{y - y_0}{b} = \frac{z - z_0}{c} \quad \text{(symmetric form)}$$

Using projections to find
    the distance between a plane and a point.
    the distance between a line and a point.
    the distance between two parallel planes.
    the distance between two skew lines.

Projections generalized for an arbitrary linear space. The linear transformation $T: V \to V$ is a projection if and only if T is idempotent ($T^2 = T$).

The vector product. Elementary properties and applications to areas and volumes. The triple scalar product.

Exercises

1. Find the equation of the plane through the point $(1, 3, -4)$ perpendicular to the vector $(2, -1, 3)$.

Applications to Geometry

2. Find the equation of the plane through the points $(1, 1, -3)$, $(2, 1, 4)$, $(4, 2, 3)$.

3. Find the equation of the plane through the point $(-5, 1, 2)$ parallel to the plane $2x + 3y - z = 4$.

4. (a) Find parametric equations of the line through $(1, 3, -6)$ and $(2, -1, 7)$.
   (b) Find symmetric equations for the line of part (a).

5. Find parametric equations of the line of intersection of $2x + y - z - 3 = 0$ and $x - 3y + 2z - 2 = 0$.

6. Let L be the line through $(2, -1, 1)$ and $(3, 1, 4)$. Find the points where L pierces
   (a) the plane $x + y - 3z = 4$.
   (b) the plane $x + y - z = 5$.
   (c) the plane $x - 2y + z = 5$.

7. Given the four points $A(5, 0, -1)$, $B(8, 1, -3)$, $C(6, -2, 2)$, $D(-6, 1, -1)$, find where the line AB meets the line CD. If the lines do not meet, find the length of the shortest line segment connecting the lines.

8. Given the four points $A(3, 2, 3)$, $B(4, 1, 2)$, $C(1, 5, -1)$, $D(3, 1, 0)$, find where the line AB meets the line CD. If the lines do not meet, find the length of the shortest line segment connecting the lines.

9. (a) Sketch the plane $5x + 6z = 30$.
   (b) Sketch the plane $2x - 3y - 6 = 0$.
   (c) Find the equation of the plane which is parallel to the x-axis and contains the points $(1, 1, -2)$ and $(2, 4, 3)$.

10. Prove that the points $(3, 0, 0)$, $(10, 1, -1)$, $(4, -1, -3)$, $(5, 2, 4)$ lie on the same plane and find the equation of the plane.

11. Find k so that the following planes are perpendicular: $(2k + 7)x + 2ky + 2z - 15 = 0$, $-kx - 3y + (4k - 1)z + 1 = 0$.

12. The points $A(2, 3, 1)$ and $B(4, 4, 3)$ are given. Find the equations of the three planes which are parallel to a coordinate axis and contain the line AB. Sketch.

13. Find the equations of the three planes which project the line $x = 2 + 3t$, $y = 1 + 3t$, $z = 3 + 4t$ upon the coordinate planes.

14. (a) Find the distance from the plane $2x + 2y - z + 3 = 0$ to the point $(3, 1, 2)$.

247

Lessons in Linear Algebra

(b) Find the distance between the planes $2x + 2y - z + 3 = 0$ and $2x - 2y - z - 9 = 0$.

15. (a) Prove that the distance from the plane $ax + by + cz + d = 0$ to the point $P_0(x_0, y_0, z_0)$ is
$$\frac{|ax_0 + by_0 + cz_0 + d|}{\sqrt{a^2 + b^2 + c^2}}$$

(b) Find the distance from the origin to the plane $ax + by + cz + d = 0$.

16. Find the distance between the lines $(x - 1)/2 = (y + 1)/3 = (z - 1)/(-1)$ and $(x + 2)/2 = (y - 3)/3 = z/(-1)$.

17. Let $O(0, 0, 0)$, $A(a_1, a_2, a_3)$, $B(b_1, b_2, b_3)$ be points of three-space. Find the coordinates of the point C such that $OACB$ is a parallelogram, and thus verify that the addition of vectors component-wise is equivalent to addition by the parallelogram rule.

18. (a) Prove that $\alpha \times \beta \cdot \alpha = 0$ and $\alpha \times \beta \cdot \beta = 0$.
    (b) Prove that $\alpha \times \beta = -\beta \times \alpha$.

19. Prove $\alpha \times (\beta + \gamma) = (\alpha \times \beta) + (\alpha \times \gamma)$, and $(\beta + \gamma) \times \alpha = (\beta \times \alpha) + (\gamma \times \alpha)$.

In the following exercises, $i = (1, 0, 0), j = (0, 1, 0), k = (0, 0, 1)$.

20. (a) Simplify each cross-product of pairs of vectors from the set i, j, k. That is, show $i \times j = k, k \times j = -i$, etc.
    (b) Simplify: $i \times (j + k)$; $(i \times j) \times (j + k)$; $(i \times j) \cdot (j \times k)$; $(i + j) \times (i \cdot i)j$.

21. Show that $\|\alpha \times \beta\|^2 = \begin{vmatrix} \alpha \cdot \alpha & \alpha \cdot \beta \\ \beta \cdot \alpha & \beta \cdot \beta \end{vmatrix}$. Then compare with Theorem 16-7, part (1).

22. Let $\alpha = i + 2j - k$, $\beta = j - k$, $\gamma = 3i + k$. Simplify: $\alpha \times \beta$; $\beta \times \gamma$; $(\alpha \times \beta) \times \gamma$; $\alpha \times (\beta \times \gamma)$, $(\alpha \times \beta) \times \beta$; $\alpha \cdot (\alpha \times \beta)$; $(\alpha \cdot \beta)(\alpha \times \beta)$.

23. Let $\alpha = i - j + k, \beta = 2i + 3j - k, \gamma = -i + 2j + 2k$. Simplify: $\alpha \times \beta$; $\alpha \times \gamma$; $\beta \times \gamma$, $(\alpha \times \beta) \times \gamma$, $\alpha \times (\beta \times \gamma)$, $\alpha \cdot (\beta \times \gamma)$; $(\alpha \times \beta) \cdot \gamma$; $(\alpha \times \beta) \times (\alpha \times \gamma)$.

24. Prove that if the sum of three vectors is zero, then the cross-product of any pair is equal to plus or minus the cross product of any other pair.

Applications to Geometry

25. (a) Prove that the area of the triangle formed by (x_1, y_1), (x_2, y_2), (x_3, y_3) is equal to the absolute value of
$$\frac{1}{2}\begin{vmatrix} x_1 & y_1 & 1 \\ x_2 & y_2 & 1 \\ x_3 & y_3 & 1 \end{vmatrix}$$
(b) Find the area of the triangle with vertices $(3, 1)$, $(2, -1)$, $(5, 4)$.

26. Find the volume of the parallelepiped determined by the vectors $\alpha = (2, 1, -4)$, $\beta = (3, 1, 2)$, $\gamma = (-1, 4, 2)$.

27. Prove that the volume of the tetrahedron with vertices $P_1(x_1, y_1, z_1)$, ..., $P_4(x_4, y_4, z_4)$ is equal to the absolute value of
$$\frac{1}{6}\begin{vmatrix} x_1 & y_1 & z_1 & 1 \\ x_2 & y_2 & z_2 & 1 \\ x_3 & y_3 & z_3 & 1 \\ x_4 & y_4 & z_4 & 1 \end{vmatrix}$$

Remarks on Exercises

1. $2x - y + 3z + 13 = 0$.
2. $7x - 15y - z + 5 = 0$.
3. $2x + 3y - z + 9 = 0$.
4. (a) $x = 1 + t$
   $y = 3 - 4t$
   $z = -6 + 13t$.
   (b) $\dfrac{x-1}{1} = \dfrac{y-3}{-4} = \dfrac{z+6}{13}$.
5. The vectors $(2, 1, -1)$ and $(1, -3, 2)$ are respective normals to the planes, so the direction of the line of intersection is given by the vector $(1, 5, 7)$, which is perpendicular to each normal. One point on the intersection line is $(0, -8, -11)$, so parametric equations (not unique) for the line are
$$(x, y, z) = (0, -8, -11) + t(1, 5, 7)$$
6. The line L has equations $x = 2 + t$, $y = -1 + 2t$, $z = 1 + 3t$.
   (a) L meets the plane $x + y - 3z = 4$ when $2 + t - 1 + 2t - 3 - 9t = 4$ or when $t = -1$. This gives the point $(1, -3, -2)$.
   (b) L meets the plane $x + y - z = 5$ when $0 = 5$. That is, L does not meet the plane.
   (c) L lies in the plane $x - 2y + z = 5$.
7. The equations of the lines are the following
   AB: $x = 5 + 3t$, $y = t$, $z = -1 - 2t$

Lessons in Linear Algebra

CD: $x = 6 + 12s$, $y = -2 - 3s$, $z = 2 + 3s$
$t = -1$ and $s = -(1/3)$ give the intersection point $(2, -1, 1)$.

8. The lines have the following equations:
    AB: $x = 3 + t$, $y = 2 - t$, $z = 3 - t$
    CD: $x = 1 + 2s$, $y = 5 - 4s$, $z = -1 + s$
    The lines do not meet. A common normal is $N = (5, 3, 2)$. The projection of $\vec{AC} = (-2, 3, -4)$ upon N is $-(9/38)(5, 3, 2)$, so the length of the shortest line segment connecting AB to CD is $\|N\| = 9/\sqrt{38}$.

9. (a) See Figure 19-13.

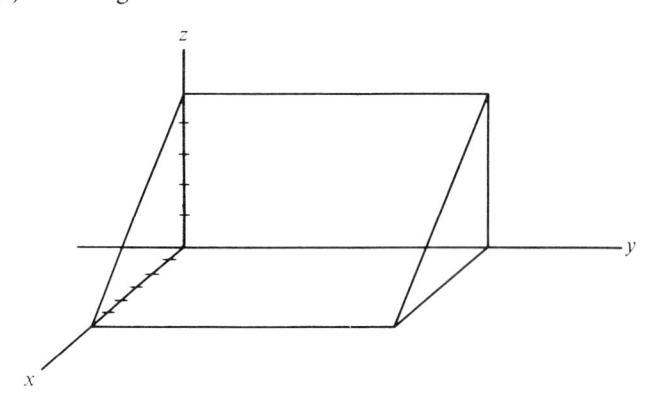

FIGURE 19-13

(b) See Figure 19-14.

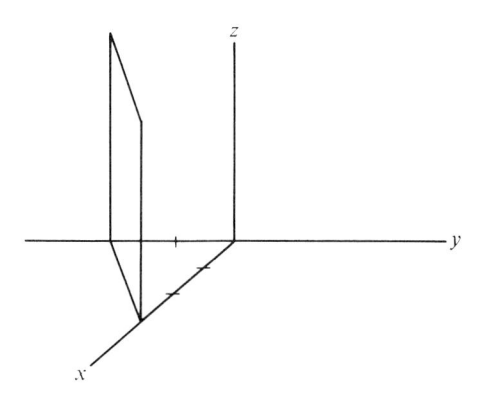

FIGURE 19-14

(c) $5y - 3z = 11$.

Applications to Geometry

10. $x - 5y + 2z = 3$.

11. The planes are perpendicular if and only if the normal vectors $(2k + 7, 2k, 2)$ and $(-k, -3, 4k - 1)$ are perpendicular. This gives $k = -1/2$ or $k = -2$.

12. $2y - z = 5$; $x - z = 1$; $x - 2y = -4$. See Figure 19-15.

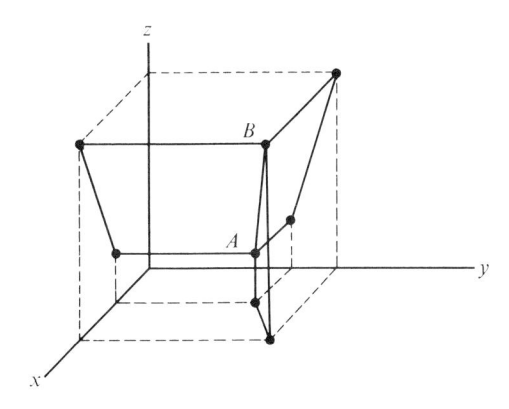

FIGURE 19-15

13. $4y - 3z = -5$; $4x - 3z = -1$; $x - y = 1$.

14. (a) 3.   (b) 4.

15. (a) Choose any point $P_1(x_1, y_1, z_1)$ on the plane and project $\overrightarrow{P_1P_0}$ upon the normal vector $N = (a, b, c)$. This gives
$$\frac{a(x_0 - x_1) + b(y_0 - y_1) + c(z_0 - z_1)}{a^2 + b^2 + c^2}(a, b, c)$$
$$= \frac{ax_0 + by_0 + cz_0 + d}{a^2 + b^2 + c^2}(a, b, c)$$
since P_1 satisfies the equation of the plane. The length of this last vector is $\dfrac{|ax_0 + by_0 + cz_0 + d|}{\sqrt{a^2 + b^2 + c^2}}$.

   (b) $\dfrac{|d|}{\sqrt{a^2 + b^2 + c^2}}$.

16. The lines are parallel. One plane normal to each line is $2x + 3y - z = 0$. Choose points $A(1, -1, 1)$, $B(-2, 3, 0)$ on the two lines and find the length of the projection of $\overrightarrow{AB}$ upon the plane $2x + 3y - z = 0$. With $N = (2, 3, -1)$ we have

251

Lessons in Linear Algebra

$$\vec{AB} - \frac{\vec{AB} \cdot N}{N \cdot N} N = \left(-4, \frac{5}{2}, -\frac{1}{2}\right)$$

and the required distance is

$$\sqrt{16 + \frac{25}{4} + \frac{1}{4}} = \frac{3}{2} \sqrt{10}$$

17. The equation of the line AC is $(x, y, z) = (a_1, a_2, a_3) + t(b_1, b_2, b_3)$, and the equation of line BC is $(x, y, z) = (b_1, b_2, b_3) + s(a_1, a_2, a_3)$. They meet when $s = t = 1$, at the point $(a_1 + b_1, a_2 + b_2, a_3 + b_3)$.

18. (a) Let $\alpha = (a_1, a_2, a_3)$, $\beta = (b_1, b_2, b_3)$. Then

$$\alpha \times \beta \cdot \alpha = \left(\begin{vmatrix} a_2 & a_3 \\ b_2 & b_3 \end{vmatrix}, \begin{vmatrix} a_3 & a_1 \\ b_3 & b_1 \end{vmatrix}, \begin{vmatrix} a_1 & a_2 \\ b_1 & b_2 \end{vmatrix} \right) \cdot (a_1 \quad a_2 \quad a_3)$$

$$= \begin{vmatrix} a_1 & a_2 & a_3 \\ a_1 & a_2 & a_3 \\ b_1 & b_2 & b_3 \end{vmatrix} = 0$$

19. Let $\alpha = (a_1, a_2, a_3)$, etc. Then

$$\alpha \times (\beta + \gamma)$$

$$= \left(\begin{vmatrix} a_2 & a_3 \\ b_2 + c_2 & b_3 + c_3 \end{vmatrix}, \begin{vmatrix} a_3 & a_1 \\ b_3 + c_3 & b_1 + c_1 \end{vmatrix}, \begin{vmatrix} a_1 & a_2 \\ b_1 + c_1 & b_2 + c_2 \end{vmatrix} \right)$$

$$= \left(\begin{vmatrix} a_2 & a_3 \\ b_2 & b_3 \end{vmatrix} + \begin{vmatrix} a_2 & a_3 \\ c_2 & c_3 \end{vmatrix}, \text{etc.} \right)$$

$$= (\alpha \times \beta) + (\alpha \times \gamma)$$

Then

$$(\beta + \gamma) \times \alpha = -\alpha \times (\beta + \gamma)$$
$$= (-\alpha \times \beta) + (-\alpha \times \gamma)$$
$$= (\beta \times \alpha) + (\gamma \times \alpha)$$

20. (a) $i \times j = k$, $j \times k = i$, $k \times i = j$, $j \times i = -k$, $k \times j = -i$, $i \times k = -j$.
    (b) $i \times (j + k) = k - j$; $(i \times j) \times (j + k) = -i$; $(i \times j) \cdot (j \times k) = 0$; $(i + j) \times (i \cdot i)j = k$.

21. $\|\alpha \times \beta\|^2 = \begin{vmatrix} a_2 & a_3 \\ b_2 & b_3 \end{vmatrix}^2 + \begin{vmatrix} a_3 & a_1 \\ b_3 & b_1 \end{vmatrix}^2 + \begin{vmatrix} a_1 & a_2 \\ b_1 & b_2 \end{vmatrix}^2$. Show that this is equal to $(\alpha \cdot \alpha)(\beta \cdot \beta) - (\alpha \cdot \beta)^2$.

Applications to Geometry

22. $\alpha \times \beta = -i + j + k$; $\beta \times \gamma = i - 3j - 3k$; $(\alpha \times \beta) \times \gamma = i + 4j - 3k$; $\alpha \times (\beta \times \gamma) = -9i + 2j - 5k$; $(\alpha \times \beta) \times \beta = -2i - j - k$; $\alpha \cdot (\alpha \times \beta) = 0$; $(\alpha \cdot \beta)(\alpha \times \beta) = 3(\alpha \times \beta) = -3i + 3j + 3k$.

23. $\alpha \times \beta = -2i + 3j + 5k$; $\alpha \times \gamma = -4i - 3j + k$; $\beta \times \gamma = 8i - 3j + 7k$; $(\alpha \times \beta) \times \gamma = -4i - j - k$; $\alpha \times (\beta \times \gamma) = -4i + j + 5k$; $\alpha \cdot (\beta \times \gamma) = (\alpha \times \beta) \cdot \gamma = 18$; $(\alpha \times \beta) \times (\alpha \times \gamma) = (18i - 18j + 18k)$.

24. If $\alpha + \beta + \gamma = 0$, then $\alpha \times \beta = \alpha \times (-\alpha - \gamma) = (\alpha + \gamma) \times \alpha = \gamma \times \alpha$, etc.

25. (a) The area is one-half the area of the parallelogram formed by the vectors $(x_2 - x_1, y_2 - y_1)$ and $(x_3 - x_1, y_3 - y_1)$. By Theorem 16-7, part 2, this area is the absolute value of

$$\begin{vmatrix} x_2 - x_1 & y_2 - y_1 \\ x_3 - x_1 & y_3 - y_1 \end{vmatrix}$$

But

$$\begin{vmatrix} x_1 & y_1 & 1 \\ x_2 & y_2 & 1 \\ x_3 & y_3 & 1 \end{vmatrix} = \begin{vmatrix} x_1 & y_1 & 1 \\ x_2 - x_1 & y_2 - y_1 & 0 \\ x_3 - x_1 & y_3 - y_1 & 0 \end{vmatrix} = \begin{vmatrix} x_2 - x_1 & y_2 - y_1 \\ x_3 - x_1 & y_3 - y_1 \end{vmatrix}$$

(b) $\dfrac{1}{2} \begin{vmatrix} 3 & 1 & 1 \\ 2 & -1 & 1 \\ 5 & 4 & 1 \end{vmatrix} = \dfrac{1}{2}.$

26. $|\alpha \times \beta \cdot \gamma| = 72.$

27.
$$\begin{vmatrix} x_1 & y_1 & z_1 & 1 \\ x_2 & y_2 & z_2 & 1 \\ x_3 & y_3 & z_3 & 1 \\ x_4 & y_4 & z_4 & 1 \end{vmatrix} = \begin{vmatrix} x_1 - x_4 & y_1 - y_4 & z_1 - z_4 & 0 \\ x_2 - x_4 & y_2 - y_4 & z_2 - z_4 & 0 \\ x_3 - x_4 & y_3 - y_4 & z_3 - z_4 & 0 \\ x_4 & y_4 & z_4 & 1 \end{vmatrix}$$

$$= \begin{vmatrix} x_1 - x_4 & y_1 - y_4 & z_1 - z_4 \\ x_2 - x_4 & y_2 - y_4 & z_2 - z_4 \\ x_3 - x_4 & y_3 - y_4 & z_3 - z_4 \end{vmatrix}$$

This is numerically equal to the volume of the parallelepiped formed by the vectors $\overrightarrow{P_4P_1}$, $\overrightarrow{P_4P_2}$, $\overrightarrow{P_4P_3}$. The volume of the tetrahedron, being one-third the base area times the altitude, is equal numerically to one-sixth of the given fourth-order determinant.

20

QUADRATIC FORMS AND QUADRIC SURFACES

In this lesson we use symmetric matrices to represent quadratic forms and we learn how to interpret the form by simplifying its matrix. The fact that any real symmetric matrix is "orthogonally" similar to a diagonal matrix is the main tool used in interpreting quadratic forms.

The interpretation of quadratic forms in two-space leads to the three conic sections, and the interpretation in three-space leads to the seventeen quadric surfaces. We do not classify these surfaces completely, but a number of examples are given to show in detail how such a classification would be made.

In the optional part of Lesson 10 we introduced the Hermitian matrix. We now restate some of the definitions and results for use in this lesson.

If $A = [a_{ij}]$ is a complex matrix, then $\bar{A} = [\bar{a}_{ij}]$ is the matrix obtained by replacing each element of A with its complex conjugate. It is easy to verify rules such as $\overline{A + B} = \bar{A} + \bar{B}$ and $\overline{AB} = \bar{A}\bar{B}$. The matrix $\bar{A}^T$ (or $\overline{A^T}$) is denoted A^*, and a matrix A for which $A^* = A$ is called *Hermitian*. It follows that a real Hermitian matrix is symmetric, and we have rules such as $(AB)^* = B^*A^*$, $A^{**} = A$, $(A + B)^* = A^* + B^*$, etc.

20-1 Theorem The eigenvalues of a Hermitian matrix are real.

Proof: Let A be an n by n Hermitian matrix, and let $Y = (y_1, \ldots, y_n)$ be any complex vector. Then the 1 by 1 matrix YAY^* is real, because

$$\overline{YAY^*} = YAY^* \quad (YAY^* \text{ is 1 by 1})$$
$$= YA^*Y^* \quad ((MN)^* = N^*M^*)$$
$$= YAY^* \quad (A \text{ is Hermitian})$$

and of course any self-conjugate complex number is real.

Now, suppose $\bar{Y}$ is an eigenvector, with $AY^* = kY^*$. Then $YAY^* = kYY^*$ and $k = YAY^*/YY^*$, a real number. ∎

Quadratic Forms and Quadric Surfaces

The eigenvalues of a real symmetric matrix are real.

20-2 Corollary

Proof: A real symmetric matrix is Hermitian. ∎

Example 1. Let

$$A = \begin{bmatrix} 1 & i \\ -i & 2 \end{bmatrix}, \quad B = \begin{bmatrix} 1 & 2 \\ 2 & 4 \end{bmatrix}$$

Then A and B are both Hermitian, and B is symmetric.

$$|A - xI| = \begin{vmatrix} 1-x & i \\ -i & 2-x \end{vmatrix} = x^2 - 3x + 1$$

and the eigenvalues of A are $(3 \pm \sqrt{5})/2$.

Similarly, the characteristic polynomial of B is $x^2 - 5x$ and the eigenvalues are $k_1 = 0$, $k_2 = 5$. The corresponding eigenvectors $X_1 = (2, -1)$ and $X_2 = (1, 2)$ are orthogonal, a fact that we now generalize.

Eigenvectors corresponding to distinct eigenvalues of a real symmetric matrix are orthogonal.

20-3 Theorem

Proof: Let k_1, k_2 be distinct eigenvalues of the real symmetric matrix A, and let X_1, X_2 be corresponding eigenvectors. Transposing $AX_1^T = k_1 X_1^T$ gives $X_1 A = k_1 X_1$; then $X_1 A X_2^T = k_1 X_1 X_2^T$. On the other hand, $AX_2^T = k_2 X_2^T$ gives $X_1 A X_2^T = k_2 X_1 X_2^T$. Therefore, $k_1 X_1 X_2^T = k_2 X_1 X_2^T$, and $(k_1 - k_2) X_1 X_2^T = 0$ implies $X_1 X_2^T = X_1 \cdot X_2 = 0$. ∎

If the n by n real symmetric matrix A has n distinct eigenvalues, Theorem 20-3 shows that A has an orthonormal set of n eigenvectors. We now prove a stronger result, that A has n orthonormal eigenvectors even if the eigenvalues are not distinct.

Let A be an nth order real symmetric matrix. Then A has n (real) orthonormal eigenvectors.

20-4 Theorem

Proof: We use mathematical induction. For $n = 1$ the result is trivial. Now from the induction hypothesis that a real symmetric matrix of order $n - 1$ has $n - 1$ orthonormal eigenvectors, we will prove that a real symmetric matrix of order n has n orthonormal eigenvectors.

Lessons in Linear Algebra

Let $T: R^n \to R^n$ be the transformation which sends $X = (x_1, \ldots, x_n)$ into XA, let X_1 be a unit eigenvector corresponding to the eigenvalue k_1, and let S be the one-dimensional subspace spanned by X_1. Extend X_1 to the orthonormal basis $\{X_1, X_2, \ldots, X_n\}$ of R^n, and note that $\{X_2, \ldots, X_n\}$ is a basis of $S^\perp$.

The transformation T sends $S^\perp$ into $S^\perp$, because for $i = 2, \ldots, n$, $X_i A X_1^T = k_1 X_i X_1^T = k_1 X_i \cdot X_1 = 0$. Restricted to the space $S^\perp$, the transformation T has matrix B (of order $n-1$) relative to the basis $\{X_2, \ldots, X_n\}$. The matrix

$$M = \begin{bmatrix} k_1 & 0 \\ 0 & B \end{bmatrix}$$

represents the unrestricted transformation T relative to the basis $\{X_1, \ldots, X_n\}$. Since A represents T relative to the standard basis, we have $M = P^{-1}AP$ where P is the transition matrix from one orthonormal basis to another. This means P is orthogonal and M is symmetric.

The induction hypothesis applied to the symmetric matrix B gives $n-1$ orthonormal eigenvectors of B. These correspond to $n-1$ orthonormal eigenvectors $\{Y_2, \ldots, Y_n\}$ of A, each eigenvector being an n-tuple with first entry zero. Let $Y_1 = (1, 0, \ldots, 0)$. Then the set $\{Y_1, Y_2, \ldots, Y_n\}$ is an orthonormal set of eigenvectors of M, and the set $\{Y_1 P^T, \ldots, Y_n P^T\}$ is an orthonormal set of eigenvectors of A. ∎

20-5 Corollary

If A is a real symmetric matrix, then there exists an orthogonal matrix P such that $P^{-1}AP$ is diagonal. (Alternatively we say that a real symmetric matrix is *orthogonally similar* to a diagonal matrix.)

Proof: We use the orthonormal set of eigenvectors assured by the theorem in order to construct the matrix P. Then apply Theorem 18-5. ∎

Example 2. Let

$$A = \begin{bmatrix} 3 & 1 \\ 1 & 3 \end{bmatrix}$$

The eigenvalues are $k_1 = 2$, $k_2 = 4$ and corresponding unit eigenvectors are $(1/\sqrt{2})(1, -1)$, $(1/\sqrt{2})(1, 1)$. With $P = (1/\sqrt{2})\begin{bmatrix} 1 & 1 \\ -1 & 1 \end{bmatrix}$, we have

$$P^{-1}AP = \begin{bmatrix} 2 & 0 \\ 0 & 4 \end{bmatrix}.$$

Quadratic Forms and Quadric Surfaces

Now use the same matrix A, and consider the quadratic form XAX^T, where $X = (x_1, x_2)$. Applying the coordinate change $X = YP^{-1}$ (or $Y = XP$) gives $XAX^T = YP^TAPY^T$, since $P^{-1} = P^T$. Written out this means that the change of coordinates $x_1 = (1/\sqrt{2})(y_1 + y_2)$, $x_2 = (1/\sqrt{2})(-y_1 + y_2)$ transforms the quadratic form $3x_1^2 + 2x_1x_2 + 3x_2^2$ into the simpler quadratic form $2y_1^2 + 4y_2^2$, a fact that can be checked by direct substitution.

If P is interpreted as a matrix of transition from the basis $\{(1, 0), (0, 1)\}$ to $\{(1/\sqrt{2})(1, -1), (1/\sqrt{2})(1, 1)\}$, then P represents a 45° rotation of axes. The ellipse whose old equation is $3x_1^2 + 2x_1x_2 + 3x_2^2 = 8$ has for its new equation $2y_1^2 + 4y_2^2 = 8$ or $(y_1^2/4) + (y_2^2/2) = 1$. See Figure 20-1.

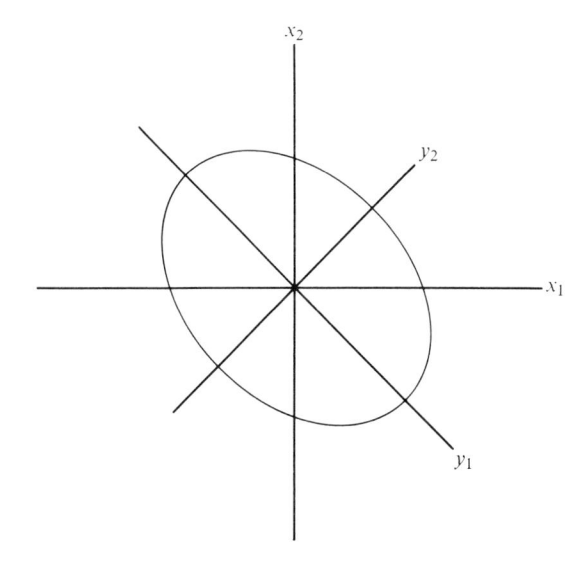

FIGURE 20-1

Example 3. Let us examine the quadric surface $5x^2 + 2y^2 + 2z^2 + 8yz + 4zx - 4xy = 36$ by simplifying the quadratic form given by the left member.

The quadratic form can be written XAX^T, where $X = (x, y, z)$, and

$$A = \begin{bmatrix} 5 & -2 & 2 \\ -2 & 2 & 4 \\ 2 & 4 & 2 \end{bmatrix}$$

Our plan is to find a matrix orthogonally similar to A.

257

Lessons in Linear Algebra

The characteristic equation of A is $|A - xI| = 0$, $-x^3 + 9x^2 - 108 = 0$, or $-(x - 6)^2(x + 3) = 0$. The eigenvalues are $6, 6, -3$. Hence, there is an orthogonal matrix P which changes the equation $XAX^T = 36$ into $X'(P^TAP)X'^T = 36$, or $6x'^2 + 6y'^2 - 3z'^2 = 36$. This is a "hyperboloid." Its graph relative to the x', y', z' coordinate system is shown in Figure 20-2.

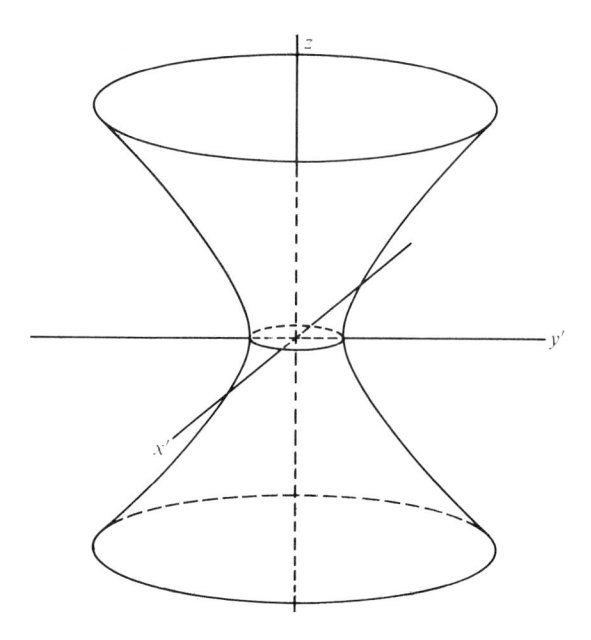

FIGURE 20-2

Examples 2 and 3 show how matrix algebra can be employed to analyze conic sections and quadric surfaces. In each example a quadratic form XAX^T was simplified by applying an orthogonal transformation to produce the "diagonalized" quadratic form $XP^{-1}APX^T$ or XP^TAPX^T.

There are three conic sections, the ellipse, the parabola, and the hyperbola. In order to determine what type of conic section is represented by the equation

$$ax^2 + 2bxy + cy^2 + dx + ey + f = 0$$

we examine the eigenvalues of the matrix

$$\begin{bmatrix} a & b \\ b & c \end{bmatrix}$$

Quadratic Forms and Quadric Surfaces

For example, if the eigenvalues k_1, k_2 are both positive, then the equation could be transformed orthogonally into $k_1 x^2 + k_2 y^2 + d'x + e'y + f' = 0$ and hence represents an ellipse.

The situation for second-degree equations in three-space is analogous. The equation

$$ax^2 + by^2 + cz^2 + 2fyz + 2gzx + 2hxy + mx + ny + pz + r = 0$$

represents one of seventeen quadric surfaces. A detailed identification may be made by examining the eigenvalues and the rank of the matrix

$$\begin{bmatrix} a & h & g \\ h & b & f \\ g & f & c \end{bmatrix}$$

A systematic treatment appears in *Linear Algebra for Undergraduates*, D. C. Murdoch; John Wiley & Sons, Inc., New York, 1957. A few illustrative examples are included in the exercises at the end of this section.

We now turn to a closer examination of the quadratic form XAX^T, where $X = (x_1, \ldots, x_n)$, and A is real symmetric.

20-6 Definition
The real quadratic forms XAX^T and XBX^T are *equivalent* if there is a nonsingular matrix P such that $PAP^T = B$. The two forms are orthogonally equivalent if there is an orthogonal matrix P such that $PAP^T = B$.

It is easy to verify that the two relations defined above are equivalence relations and it is obvious that requiring the existence of a matrix P such that $PAP^T = B$ is the same as requiring the existence of a matrix P satisfying $P^T A P = B$.

We already know from Corollary 20-5 that XAX^T is "orthogonally equivalent" to the diagonalized form $YBY^T = k_1 y_1^2 + \ldots + k_n y_n^2$, where $k_1, \ldots, k_n$ are eigenvalues of A. These eigenvalues tell us much about the range of XAX^T.

20-7 Theorem
Let A, B be two matrices such that there exists a nonsingular matrix P satisfying $PAP^T = B$. Then the quadratic forms XAX^T and YBY^T have the same range.

Proof: For a given Y, $YBY^T = YPA(YP)^T$. Hence, $XAX^T = YBY^T$ when $X = YP$, or when $Y = XP^{-1}$. ∎

Lessons in Linear Algebra

20-8
Definition

The quadratic form XAX^T is *positive definite* if $XAX^T > 0$ when $X \neq 0$. The form is *nonnegative semidefinite* if $XAX^T \geq 0$ for all X.

Example 4. The quadratic form $x_1^2 + 2x_1x_2 + 2x_2^2$ is positive definite since it can be written as the sum of two squares $(x_1 + x_2)^2 + x_2^2$ and is clearly positive unless $(x_1, x_2) = (0, 0)$.

The quadratic form $x_1^2 + x_1x_2 - x_2^2$ is "indefinite" since it takes on both positive and negative values.

The quadratic form $x_1^2 + 2x_1x_2 + x_2^2$ is nonnegative semidefinite because $(x_1 + x_2)^2 \geq 0$ for all (x_1, x_2). However, it is not positive definite, because $(x_1 + x_2)^2 = 0$ does not imply $(x_1, x_2) = (0, 0)$.

20-9
Theorem

Let A be a real symmetric matrix. The quadratic form XAX^T is positive definite if and only if the eigenvalues of A are all positive. The quadratic form is nonnegative semidefinite if and only if the eigenvalues of A are all nonnegative.

Proof: Let $k_1, \ldots, k_n$ be the eigenvalues of A. By Corollary 20-5, A is orthogonally similar to a diagonal matrix B. In other words, the given quadratic form XAX^T is orthogonally equivalent to

$$YBY^T = k_1y_1^2 + \ldots + k_ny_n^2$$

Furthermore, Theorem 20-7 assures that XAX^T and YBY^T have the same range.

Therefore, each form is positive definite if all the eigenvalues are positive, and each form is nonnegative semidefinite if all the eigenvalues are nonnegative. ∎

Example 5. Let us consider the quadratic form $XAX^T = 4x_1^2 + 3x_2^2 + 2x_3^2 - 4x_1x_2 + 4x_2x_3$, where

$$A = \begin{bmatrix} 4 & -2 & 0 \\ -2 & 3 & 2 \\ 0 & 2 & 2 \end{bmatrix}$$

The characteristic polynomial of A is $|A - xI|$, or $-x^3 + 9x^2 - 18x$. Since the eigenvalues of A are 6, 3, and 0, A is orthogonally similar to the quadratic form

$$XBX^T = 6x_1^2 + 3x_2^2$$

Quadratic Forms and Quadric Surfaces

where
$$B = \begin{bmatrix} 6 & 0 & 0 \\ 0 & 3 & 0 \\ 0 & 0 & 0 \end{bmatrix}$$

and it is clear that XAX^T is nonnegative semidefinite.

The above casts light on the quadric surface $4x^2 + 3y^2 + 2z^2 - 4xy + 4yz = 12$. An orthogonal change of coordinates presents the surface as $6x^2 + 3y^2 = 12$, an "elliptic cylinder." See Figure 20-3.

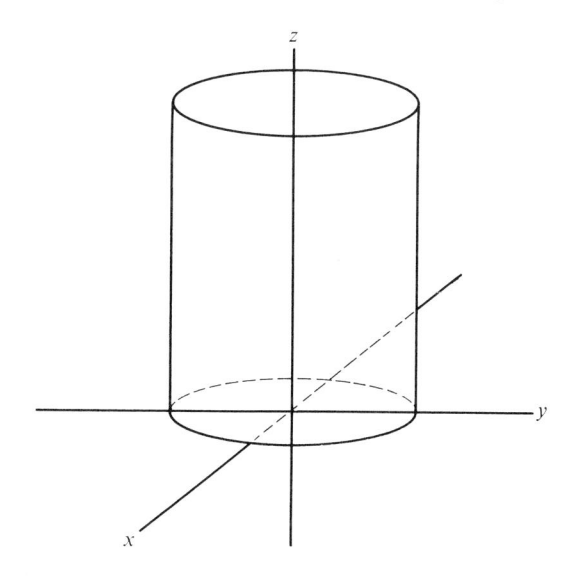

FIGURE 20-3

That the range of XAX^T is nonnegative can be seen more easily, if we "complete some squares." For

$$\begin{aligned} XAX^T &= 4x_1^2 + 3x_2^2 + 2x_3^2 - 4x_1x_2 + 4x_2x_3 \\ &= (4x_1^2 - 4x_1x_2 + x_2^2) + 2x_2^2 + 4x_2x_3 + 2x_3^2 \\ &= (2x_1 - x_2)^2 + 2(x_2 + x_3)^2 \end{aligned}$$

and clearly $XAX^T \geq 0$ for all X. It is also clear that there is a nonorthogonal transformation of coordinates which changes the quadratic form into $YBY^T = y_1^2 + y_2^2$, a much simpler form which has the same range as XAX^T. The form $YBY^T = y_1^2 + y_2^2$ is called the *real canonical form* of the given quadratic form.

Lessons in Linear Algebra

Example 6. We find the range of $x^2 + 2y^2 + 17z^2 + 2xy - 4xz + 2yz$, by completing squares.

$$(x^2 + 2xy - 4xz) + 2y^2 + 17z^2 + 2yz$$
$$= (x + y - 2z)^2 + y^2 + 13z^2 + 6yz$$
$$= (x + y - 2z)^2 + (y + 3z)^2 + 4z^2$$

The form has the same range as $x^2 + y^2 + z^2$ and is positive definite.

We conclude the section by showing that any real quadratic form is equivalent to a "diagonalized" form XDX^T with only elements 1, -1, or 0 along the diagonal of the matrix D.

Example 7. Let A be a symmetric matrix with eigenvalues 4, 3, -16, 0. Then there is an orthogonal matrix P such that

$$P^T A P = \begin{bmatrix} 4 & 0 & 0 & 0 \\ 0 & 3 & 0 & 0 \\ 0 & 0 & -16 & 0 \\ 0 & 0 & 0 & 0 \end{bmatrix}$$

Now, let

$$Q = \begin{bmatrix} \frac{1}{2} & 0 & 0 & 0 \\ 0 & \frac{1}{\sqrt{3}} & 0 & 0 \\ 0 & 0 & \frac{1}{4} & 0 \\ 0 & 0 & 0 & 0 \end{bmatrix}$$

and we have

$$Q^T P^T A P Q = \begin{bmatrix} 1 & 0 & 0 & 0 \\ 0 & 1 & 0 & 0 \\ 0 & 0 & -1 & 0 \\ 0 & 0 & 0 & 0 \end{bmatrix}$$

It follows that the quadratic form XAX^T is equivalent to the form $x_1^2 + x_2^2 - x_3^2$, a fact that we generalize.

20-10 Theorem

Let A be a real symmetric matrix. Then the quadratic form XAX^T is equivalent to the *real canonical form* $x_1^2 + \ldots + x_p^2 - x_{p+1}^2 \ldots - x_{p+n}^2$, where p equals the number of positive eigenvalues, and n equals the num-

Quadratic Forms and Quadric Surfaces

ber of negative eigenvalues. Moreover, the real canonical form of XAX^T is unique.

Proof: The proof that the real canonical form exists is suggested clearly by Example 7. The uniqueness is harder to establish.

Let XD_1X^T and YD_2Y^T be two real canonical forms of the same quadratic form. Then D_1 and D_2 are of the same rank, since $D_2 = PD_1P^T$ where P is nonsingular. Suppose

$$XD_1X^T = x_1^2 + \ldots + x_p^2 - x_{p+1}^2 \ldots$$
$$YD_2Y^T = y_1^2 + \ldots + y_q^2 - y_{q+1}^2 \ldots$$

where $q < p$ and $X = YP$. Setting each $x_{p+1} = x_{p+2} = \ldots = 0$ gives

$$x_1^2 + \ldots + x_p^2 = y_1^2 + \ldots + y_q^2 - y_{q+1}^2 - \ldots$$

Furthermore, since $q < p$ there is a nontrivial p-tuple $x_1, \ldots, x_p$ for which $y_1 = y_2 = \ldots = y_q = 0$. This leads to the contradiction $x_1^2 + \ldots + x_p^2 = -y_{q+1}^2 - \ldots$. ∎

Summary of Topics in Lesson 20

The eigenvalues of a Hermitian matrix are real and as a corollary the eigenvalues of a real symmetric matrix are real.

For a real symmetric matrix, eigenvectors corresponding to distinct eigenvalues are orthogonal.

A real symmetric matrix of order n has an orthonormal set of n eigenvectors and is therefore orthogonally similar to a diagonal matrix.

The reduction of a quadratic form to a canonical form, or the diagonalization of a quadratic form.

The interpretation of quadratic forms in two-space and three-space in relation to conic sections and quadric surfaces.

The range of a quadratic form. Positive definite and nonnegative semidefinite quadratic forms.

The real canonical form of a quadratic form.

Exercises

1. Find the eigenvalues and eigenvectors of the Hermitian matrix

$$A = \begin{bmatrix} 4 & 1+i \\ 1-i & 3 \end{bmatrix}$$

Lessons in Linear Algebra

2. Let
$$A = \begin{bmatrix} 2 & 2 & 1 \\ 2 & -1 & -2 \\ 1 & -2 & 2 \end{bmatrix}$$

  (a) Find the eigenvalues and eigenvectors of A.
  (b) Construct an orthogonal matrix P such that $P^{-1}AP$ is diagonal.
  (c) Find a quadratic form orthogonally similar to $2x^2 - y^2 + 2z^2 + 4xy + 2xz - 4yz$.
  (d) Identify the quadric surface
  $$2x^2 - y^2 + 2z^2 + 4xy + 2xz - 4yz = -12$$
  and sketch the graph.

3. Consider the real symmetric matrix $A = \begin{bmatrix} a & b \\ b & c \end{bmatrix}$.

  (a) Explain how one may use the signs of trace (A) and $\det A$ in order to find the signs of the eigenvalues k_1, k_2 of A.
  (b) Specify conditions on a, b, c such that the conic section $ax^2 + 2bxy + cy^2 + dx + ey + f = 0$ is an ellipse; a parabola; a hyperbola.

4. Prove that orthogonal similarity of matrices is an equivalence relation.

5. Consider the quadratic form
$$XAX^T = 5x^2 + 7y^2 + 6z^2 + 4xz + 4yz$$

  (a) Find the eigenvalues of A and the range of XAX^T.
  (b) Find a set of orthonormal eigenvectors of A.
  (c) Identify and sketch the quadric surface $XAX^T = 18$ on a new set of axes. Describe the change from the old to the new axes.

6. Given
$$A = \begin{bmatrix} 2 & 0 & 2 \\ 0 & 4 & 2 \\ 2 & 2 & 3 \end{bmatrix} \quad \text{and} \quad X = (x, y, z)$$

  (a) Identify and sketch the quadric surface $XAX^T = 6$.
  (b) Describe the surface $XAX^T = -6$.
  (c) Describe the surface $XAX^T = 0$.

7. (a) Given
$$A = \begin{bmatrix} 0 & \tfrac{1}{2} & 0 \\ \tfrac{1}{2} & 0 & 0 \\ 0 & 0 & 0 \end{bmatrix}$$
find an orthogonal matrix P such that $P^{-1}AP$ is diagonal.

Quadratic Forms and Quadric Surfaces

(b) Use part (a) to transform the quadratic form XAX^T, where $X = (x, y, z)$. Then sketch the graph of the quadric surface $xy = -z$.

8. Show that if A is a singular n by n real matrix, then the quadratic form XAX^T is not positive definite.

9. The real matrix A is called *normal* if $AA^T = A^TA$.
    (a) Prove that a real symmetric matrix is normal.
    (b) Prove that a real skew-symmetric matrix is normal.
    (c) Prove that an orthogonal matrix is normal.
    (d) If $AA^T = 0$, what can be said about A?
    (e) Prove: A real upper triangular matrix A is diagonal if and only if A is normal. (One way is easy. For the other compare the ij-elements of AA^T and A^TA.)

10. Prove that the real matrix A is orthogonally similar to a diagonal matrix if and only if A is symmetric.

11. (a) Find the real canonical form of the quadratic form $5x^2 + 7y^2 + 6z^2 + 4xz + 4yz$, given that the eigenvalues of the associated matrix are all positive.
    (b) Check part (a) by completing squares.

12. Reduce $5x^2 + 2y^2 + 2z^2 + 8yz + 4xz - 4xy$ to real canonical form by completing squares. (See Example 3.)

13. Find the real canonical form of $x_1^2 + 2x_2^2 + 2x_3^2 + x_4^2 + 2x_1x_2 - 2x_1x_3 - 4x_2x_3 - 2x_2x_4 - 2x_3x_4$.

14. The *signature* of the real quadratic form XAX^T is defined as the number of positive eigenvalues minus the number of negative eigenvalues. Prove that two quadratic forms are equivalent if and only if they have the same signature and the same rank.

Optional Exercises on Complex Matrices

15. Recall that two complex vectors $X = (x_1, \ldots, x_n)$ and $Y = (y_1, \ldots, y_n)$ are orthogonal if the inner product $XY^* = 0$. Parallel Theorem 20-3 and prove that eigenvectors corresponding to distinct eigenvalues of a Hermitian matrix are orthogonal.

16. Show that a Hermitian matrix is "unitarily" similar to a real diagonal matrix. (See Theorem 20-4 and Corollary 20-5.)

17. Let A be an n by n Hermitian matrix and let $X = (x_1, \ldots, x_n)$, a complex vector. Then XAX^* is called a "complex quadratic form."

Lessons in Linear Algebra

  (a) Show XAY^* and YAX^* are conjugates and that XAX^* is real.
  (b) Show that XAX^* is "unitarily" equivalent to the form XBX^*, where B is a real diagonal matrix.
  (c) Show that XAX^* is positive definite if and only if all the eigenvalues of A are positive.
  (d) Show that XAX^* is equivalent to a form $x_1^2 + \ldots + x_p^2 - x_{p+1}^2 - \ldots - x_r^2$, where r is the rank of A and p is the number of positive eigenvalues.

18. The complex matrix A is called *normal* if $AA^* = A^*A$.
    (a) Prove that if A is Hermitian, then A is normal.
    (b) Prove that any unitary matrix is normal.
    (c) Prove that if A is normal, then A^T and $\bar{A}$ are normal.

19. (a) Show that the complex quadratic form with matrix AA^* is nonnegative semidefinite.
    (b) Show that if $AA^* = 0$, then $A = 0$.

Remarks on Exercises

1. $k_1 = 2$, $X_1 = (1 + i, -2)$; $k_2 = 5$, $X_2 = (1 + i, 1)$.

2. (a) eigenvalues 3, 3, -3; eigenvectors $(1, 0, 1)$, $(1, 1, -1)$, $(-1, 2, 1)$.
   (b) Normalize the eigenvectors from part (a) and use these to form the columns of P. The matrix P is not unique.
   (c) $3x^2 + 3y^2 - 3z^2$.
   (d) From (c), an orthogonal change of coordinates represents the surface as $x^2 + y^2 - z^2 = -4$. The graph relative to the new system is shown in Figure 20-4. The quadric surface is a "hyperboloid with two sheets."

3. (a) $k_1 + k_2 = $ trace (A) and $k_1 k_2 = \det A$.
   (b) $b^2 - ac < 0$, ellipse; $b^2 - ac = 0$, parabola; $b^2 - ac > 0$, hyperbola. This includes some unusual conics such as the imaginary ellipse $x_2 + 4y^2 = -5$.

5. (a) The eigenvalues are 3, 6, 9 and the range of XAX^T is the set of nonnegative real numbers.
   (b) $(1/3)(2, 1, -2)$, $(1/3)(2, -2, 1)$, $(1/3)(1, 2, 2)$.
   (c) Use the eigenvectors in (b) as columns of the matrix P. The transformation of coordinates $X = YP^T$ changes $XAX^T = 18$ into $3x'^2 + 6y'^2 + 9z'^2 = 18$, or (drop the primes) $x^2 + 2y^2 +$

Quadratic Forms and Quadric Surfaces

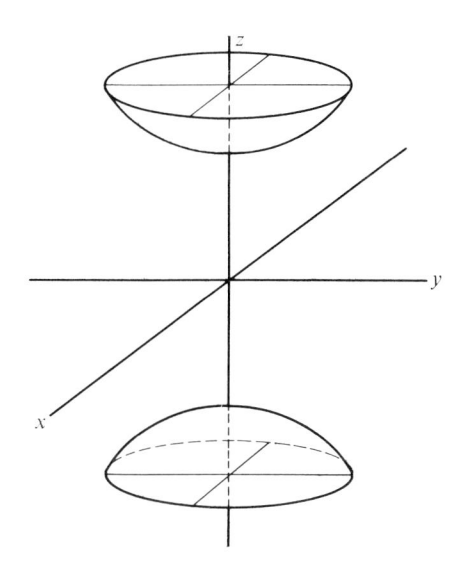

FIGURE 20-4

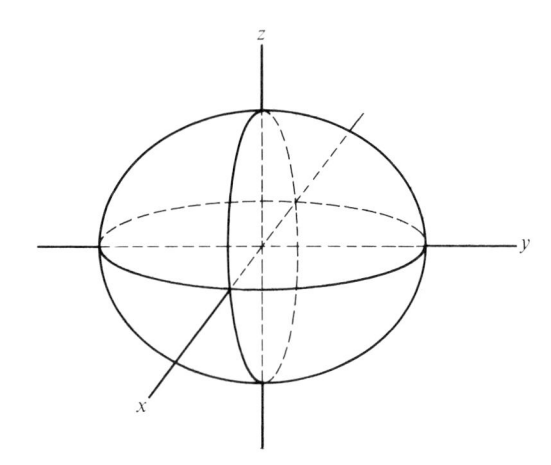

FIGURE 20-5

$3z^2 = 6$, easily recognized as an ellipsoid (shown in Figure 20-5).

6. The eigenvalues of A are 6, 3, 0. Therefore, a transformation of coordinates changes XAX^T into $6x^2 + 3y^2$. See Figure 20-6.

Lessons in Linear Algebra

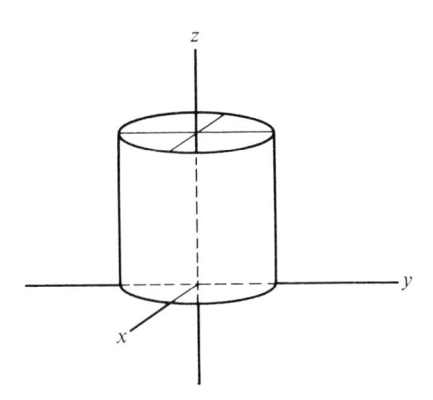

FIGURE 20-6

(a) $2x^2 + y^2 = 2$ is an elliptic cylinder.
(b) $2x^2 + y^2 = -2$ is an imaginary elliptic cylinder.
(c) $2x^2 + y^2 = 0$ is simply the z-axis, but can be viewed as a degenerate elliptic cylinder.

7. The eigenvalues of A are $1/2, -1/2, 0$.

$$P = \begin{bmatrix} \frac{1}{\sqrt{2}} & -\frac{1}{\sqrt{2}} & 0 \\ \frac{1}{\sqrt{2}} & \frac{1}{\sqrt{2}} & 0 \\ 0 & 0 & 1 \end{bmatrix}$$

(b) The original form $XAX^T = xy$ is changed (by P) into $(1/2)x^2 - (1/2)y^2$. The graph of the hyperbolic paraboloid $x^2 - y^2 = -2z$ is shown in Figure 20-7.

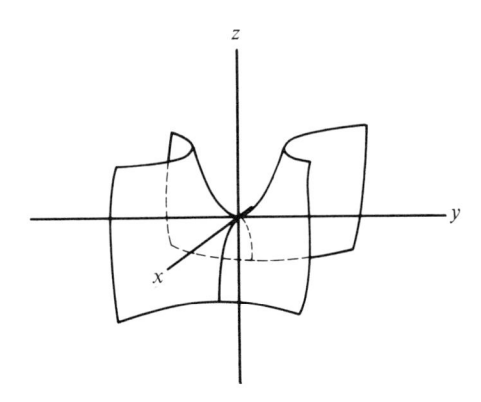

FIGURE 20-7

Quadratic Forms and Quadric Surfaces

8. If A is singular at least one eigenvalue is 0, and there is a nonzero X for which $AX^T = 0$ and $XAX^T = 0$. Also, the real canonical form of XAX^T is $x_1^2 + \ldots + x_r^2$ where the rank r is less than n.

9. (d) A must be zero, because $AA^T = 0$ implies that each row of A has length zero.
   (e) Let A be real and upper triangular. If A is diagonal it is trivial to observe that A is normal. Conversely, let A be normal (and triangular). Equating the ij-elements of AA^T and A^TA gives

$$\sum_{k=1}^{n} a_{ik}a_{jk} = \sum_{h=1}^{n} a_{hi}a_{hj}$$

Remember that $i > j$ implies $a_{ij} = 0$. For $i = j = 1$, we have $a_{11}^2 + \ldots + a_{1n}^2 = a_{11}^2$, and hence $a_{12} = a_{13} = \ldots = a_{1n} = 0$. For $i = j = 2$, we have $a_{22}^2 + \ldots + a_{2n}^2 = a_{12}^2 + a_{22}^2$, and hence $a_{23} = a_{24} = \ldots = a_{2n} = 0$. We use induction. If all the off-diagonal elements in the first $r - 1$ rows are zero, then

$$a_{rr}^2 + \ldots + a_{rn}^2 = a_{1r}^2 + \ldots + a_{rr}^2$$

gives

$$a_{r,r+1}^2 + \ldots + a_{rn}^2 = 0$$

and all of the off-diagonal elements of row r are zero.

10. If A is orthogonally similar to the diagonal matrix D, let $A = PDP^{-1}$ and show that $A = A^T$. The converse is Corollary 20-5.

11. (a) $x^2 + y^2 + z^2$.

12. $x^2 + y^2 - z^2$.

13. $(x_1 + x_2 - x_3)^2 + (x_2 - x_3 - x_4)^2 + (x_3 - x_4)^2 - (x_3 + x_4)^2$, or

$$y_1^2 + y_2^2 + y_3^2 - y_4^2$$

14. By Theorem 20-10, two quadratic forms are equivalent if their matrices have the same number p of positive eigenvalues and the same number n of negative eigenvalues. Now $p + n = r$, the rank. The signature $s = p - n$, and hence $p = (1/2)(r + s)$, and $n = (1/2)(r - s)$.

15. Let k_1, k_2 be distinct eigenvalues of the Hermitian matrix A, with X_1, X_2 as corresponding eigenvectors. Since $A^* = A$ and $\bar{A} = A^T$, transposing $AX_1^T = k_1 X_1^T$ gives $X_1 \bar{A} = k_1 X_1$; then $X_1 \bar{A} X_2^* = k_1 X_1 X_2^*$. On the other hand, $AX_2^T = k_2 X_2^T$ gives $\bar{A} X_2^* = k_2 X_2^*$ and $X_1 \bar{A} X_2^* = k_2 X_1 X_2^*$. Then $(k_1 - k_2) X_1 X_2^* = 0$ shows X_1 and X_2 are orthogonal.

Lessons in Linear Algebra

16. Parallel Theorem 20-4 and Corollary 20-5.

17. (a) $\overline{XAY^*} = (XAY^*)^* = YA^*X^* = YAX^*$.
    (b) Use Exercise 16.
    (c) Use Exercise 16.
    (d) See the proof of Theorem 20-10.

18. (c) Let $AA^* = A^*A$. Then A^T is normal if and only if
$$A^T(A^T)^* = (A^T)^*A^T$$
or
$$A^T(A^*)^T = (A^*)^T A^T$$
or
$$(A^*A)^T = (AA^*)^T$$

Therefore, A^T is normal. The proof that $\bar{A}$ is normal is similar.

19. (a) Consider the range of XAA^*X^* in the form $(XA)(XA)^*$.

BIBLIOGRAPHY

Apostol, Tom M., *Calculus*. vol. 2. 2nd ed. Waltham, Massachusetts: Blaisdell Publishing Company, 1969.

    An excellent introduction to linear algebra and multivariable calculus. For the gifted student.

Bell, E. T., *Men of Mathematics*. New York: Simon and Schuster, 1965.

    A readable account of the lives and accomplishments of great mathematicians.

Birkhoff, Garrett, and Saunders Maclane, *A Survey of Modern Algebra*. 3rd ed. New York: The Macmillan Company, 1965.

    A treatment of linear algebra is included in this text on modern algebra.

Finkbeiner, Daniel T., II, *Introduction to Matrices and Linear Transformations*. 2nd ed. San Francisco: W. H. Freeman and Company, 1966.

Halmos, Paul R., *Finite-Dimensional Vector Spaces*. 2nd ed. New York: Van Nostrand Reinhold Company, 1958.

    Recommended for its clarity and depth. A source of good exercises.

Jacobson, Nathan, *Lectures in Abstract Algebra*. vol. 2. Princeton: D. Van Nostrand Company, Inc., 1960.

Lange, L. H., *Elementary Linear Algebra*. New York: John Wiley and Sons, Inc., 1968.

Murdoch, D. C., *Linear Algebra for Undergraduates*. New York: John Wiley and Sons, Inc., 1957.

    Includes applications of matrix theory to geometry.

Bibliography

Nering, Evar D., *Linear Algebra and Matrix Theory*. New York: John Wiley and Sons, Inc., 1963.

Noble, Ben, *Applied Linear Algebra*. Englewood Cliffs, New Jersey: Prentice-Hall, Inc., 1969.

INDEX

Addition of geometric vectors, 7–9
Addition of linear transformations, 161
Adjoint of a matrix, 186
Areas in terms of determinants, 187–89

Basis:
   change, 214
   defined, 31
   extending a linearly independent set to a basis, 41
   standard, 137
Bessel, Friedrich Wilhelm, 89
Bessel inequality, 89 (Ex. 9)
Bilinear form, 108 (Ex. 3)

Cancellation law, 96
Cayley, Arthur, 224
Cayley-Hamilton theorem, 224
Change of basis, 214
Characteristic equation, 203
Characteristic polynomial, 203
Characteristic root (*see* Eigenvalue)
Characteristic vector (*see* Eigenvector)
Cofactor, 175
Column space of a matrix, 124
Complex linear space, 18
Complex numbers, basic facts, 9–10
Composition of linear transformations, 139
Conformable matrices, 94
Congruent matrices, 228 (Ex. 2b)
Coordinates relative to a basis, 149
Cramer, Gabriel, 197

Cramer's rule, 197 (Ex. 19)
Cross-product (*see* Vector product)
Determinant:
   cofactor, 175
   definition, 170
   differentiation, 181 (Ex. 21)
   Jacobian, 190
   of a linear transformation, 187
   a product of eigenvalues, 228
   of the product of two matrices, 185
   properties, 172
   test for invertibility of a matrix, 185
   test for linear independence, 184
   Vandermonde, 181 (Ex. 19)
   volume of tetrahedron, 249
Diagonal matrix, 206
Diagonalized quadratic form, 258
Dimension:
   defined, 33
   of subspaces and that of their sum and intersection, 44
Direct sum, 46, 242
Direction numbers, 235
Distance:
   between a line and point, 240
   between parallel planes, 241
   between a plane and point, 239
   between skew-lines, 241
   between vectors, 64
Dot product, 9, 61

Eigenfunction, 201

Index

Eigenspace, 204
Eigenvalue:
   definition, 200, 203
   Hermitian matrix, 254
   of an improper orthogonal matrix, 229 (Ex. 16)
   problem, 202
   of a proper orthogonal matrix, 229 (Ex. 16)
   of a real symmetric matrix, 255
Eigenvector:
   definition, 200, 203
   of a real symmetric matrix, 255
Elementary matrix, 119
Elementary operation, 50, 119
Equivalence relation, 55
Euclidean space (real inner product space), 60, 61

Fourier, Joseph, 71
Fourier series, 71
Function (*see* Mapping)
Function space, 18

Gauss, Carl Friedrich, 51
Gauss-Jordan method, 51
Generalized associative law, 99 (Ex. 4)
Generate, 30
Geometric vectors, 7–9
Gram-Schmidt process, 75

Halmos, Paul, 227
Hamilton, William Rowan, 224
Hamilton-Cayley theorem, 224
Hermite, Charles, 53
Hermite normal form (reduced echelon form), 51–56
   defined, 53
Hermitian matrix, 111, 254
Hilbert, David, 67
Hilbert space, 67
Homogeneous linear problem, 157

Idempotent, 234, 242
Identity mapping, 4
Identity matrix, 97
Image space of a linear transformation, 146
Improper orthogonal matrix, 191
Incidence matrix, 100 (Ex. 12)

Indexed set, 2
Inner product, 60–66
   defined, 60–61
Inner product space, 60
Inverse image, 3
Inverse of a linear transformation, 149
Inverse of a matrix, 118
Inversion, 170
Invertibility tests, 150
Invertible matrix, 119
   tested by determinants, 187
Isometry, 190
Isomorphism, 151

Jacobi, Carl Gustav, 190
Jacobian, 190
Jacobson, Nathan, 43
Jordan, Camille, 51
Jordan normal form, 226

Kernel of a linear transformation, 146
Kirchhoff, Gustav Robert, 56
Kronecker, Leopold, 176
Kronecker delta, 176

$L(A)$, the set of all linear combinations of A, 30
Left-inverse, right-inverse, 4
Legendre, Adrien Marie, 71
Legendre polynomials, 71
Length (norm) of a vector, 63
Line:
   parametric equations, 236
   symmetric equations, 236
   vector equation, 236
Linear combination:
   defined, 24
   trivial, 24
Linear dependence and linear independence:
   defined, 24
   property of an indexed set, 27
   tested by determinants, 184
Linear equations:
   consistent, inconsistent, 49
   homogeneous, 45
Linear mapping (*see* Linear transformation)

Index

Linear problem:
 defined, 157
 homogeneous, 157
 nonhomogeneous, 157
Linear relation, 24
Linear space (vector space):
 basis, 31
 C^n of complex n-tuples, 18
 complex, 18
 defined, 15
 dimension, 33
 finite-dimensional, 33
 function space, 18
 infinite dimensional, 33
 isomorphic, 151
 n-tuple space, 17, 18
 polynomial space, 18
 R^n of real n-tuples, 17
 real, 18
 spanning set, 32
Linear transformation:
 addition, 161
 composition, 139
 definition, 131
 determinant, 187
 determined by action on a basis, 134
 eigenspace, 204
 eigenvalue, 200
 eigenvector, 200
 idempotent, 234, 242
 image space, 146
 inverse, 149
 invertibility tests, 150
 Jacobian, 190
 kernel, 146
 matrix, 134
 nullity, 146
 orthogonal, 196
 rank, 146
 scalar multiple, 161
 zero, 162
Linearity condition, 131

Mapping:
 composition, 4

Mapping: (*continued*)
 defined, 3
 domain, codomain, range, 3
 left-inverse, right-inverse, 4
 one-to-one, onto, 3, 133
Matrix:
 addition, 94
 adjoint, 186
 characteristic equation, 203
 characteristic polynomial, 203
 column space, 124
 complex, 52
 congruent, 228 (Ex. 2b)
 defined, 52, 94
 diagonal, 206
 eigenvalue, 203
 eigenvector, 203
 elementary, 119
 equality, 53, 94
 Hermitian, 111, 254
 identity, 97
 improper orthogonal, 191
 inverse, 118
 invertible, 118
 Jordan normal form, 226
 of a linear transformation, 134
 multiplication, 94
 negative, 97
 nonsingular, 118
 normal, 265 (Ex. 9)
 orthogonal, 191
 orthogonally similar, 256
 partitioned into submatrices, 219
 proper orthogonal, 191
 rank, 125
 real, 52
 reduction to diagonal form, 217
 reduction to triangular form, 220
 representing differential equations, 109 (Ex. 4)
 row-equivalent, 55
 row space, 124
 scalar multiple, 94
 similar, 217
 singular, 118
 skew-Hermitian, 111

Index

Matrix: (*continued*)
  skew-symmetric, 107
  symmetric, 107
  trace, 127 (Ex. 9)
  of transition, 214
  transpose, 106
  triangular, 206
  type, 52, 94
  unitary, 193
  use in linear systems, 50–52
  zero, 97
Matrix algebra laws, 96, 97
Maximal linearly independent set, 37
Metric space, 64
Minimal spanning set, 37
Murdoch, D. C., 259
Mutually orthogonal sets, 70

Negative of a matrix, 97
Nering, Evar D., 227
Nonhomogeneous linear problem, 157
Nonnegative semidefinite, 260
Nonsingular matrix, 118
Nontrivial linear relation, 24
Norm (length) of a vector, 63
Normal matrix, 265 (Ex. 9), 266 (Ex. 18)
Nullity of a linear transformation, 146

Orthogonal and orthonormal bases, 75
Orthogonal complement, 83
Orthogonal linear transformation, 196 (Ex. 14), 229 (Ex. 15)
Orthogonal matrix defined, 191
Orthogonal sets, 70
Orthogonal vectors, 70
Orthogonally similar matrices, 256
Orthonormal set, 70

Parametric equations of a line, 236
Partitioned matrices, 219
Plane, equation, 234
Polynomial spaces, 18
Positive definite, 260

Projection:
  defined, 242
  of one vector upon another, 73
  of three-space, 238
  of a vector on a plane, 239
  of a vector upon a subspace, 85
Proper orthogonal matrix defined, 191
Pythagorean theorem, 85

Quadratic form, 108 (Ex. 3)
  complex, 265 (Ex. 17)
  diagonalized, 258
  equivalent, 259
  nonnegative semidefinite, 260
  orthogonally equivalent, 259
  positive definite, 260
  real canonical form, 261, 262
  signature, 265 (Ex. 14)
Quadric surfaces, 254, 257–59

Rank:
  of a linear transformation, 146
  of a matrix, 125
  of the product of matrices, 125, 153 (Ex. 10d)
Real canonical form, 261, 262
Real linear space, 18
Reduced echelon form (*see* Hermite normal form), 51–56
Reduction to diagonal form, 217
Reduction to triangular form, 220
Row-equivalent matrices, 55
Row space of a matrix, 124

Scalar multiple of a linear transformation, 161
Scalar multiplication of geometric vectors, 7–9
Schwartz inequality, 62, 65
Sets, 1–2
  intersection, 2, 21
  subsets, 2, 21
  union, 2, 21
Signature, 265 (Ex. 14)
Similar matrices, 217
Singular matrix, 118
Skew-Hermitian matrix, 111
Skew-symmetric matrix, 107

Index

Span, 30
Spanning set, 32
Standard basis, 137
Submatrices, 219
Subspace:
  defined, 21
  intersection and union, 35–36
  sum, 35
Sum of subspaces, 35
Summation notation, 5–6
Symmetric equations of a line, 236
Symmetric matrix, 107
Systems of equations, 50

Tetrahedron, volume by determinant, 249
Trace of a matrix, 127 (Ex. 9)
Transpose of a matrix, 106
Triangle inequality, 63–64
Triangular matrix, 206

Triple scalar product, 244
Trivial linear combination, 24
Trivial linear relation, 24

Unit vector, 63
Unitary matrix, 193
Unitary space, 60, 65

Vandermonde's determinant, 181 (Ex. 19)
Vector:
  geometric, 7–9
  operations on geometric vectors, 7–9
  unit vector, 12
Vector product (cross-product), defined, 244
Vector space (*see* Linear space)

Zero matrix, 97
Zero transformation, 162